# The Full-Service Community School Movement

# The Full-Service Community School Movement

## Lessons from the James Adams Community School

Jeanita W. Richardson

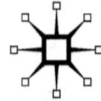

THE FULL-SERVICE COMMUNITY SCHOOL MOVEMENT
Copyright © Jeanita W. Richarson, 2009.
Softcover reprint of the hardcover 1st edition 2009

All rights reserved.

First published in 2009 by PALGRAVE MACMILLAN® in the
United States - a division of St. Martin's Press LLC, 175 Fifth Avenue,
New York, NY 10010.

Where this book is distributed in the UK, Europe and the rest of
the world, this is by Palgrave Macmillan, a division of Macmillan
Publishers Limited, registered in England, company number 785998,
of Houndmills, Basingstoke, Hampshire RG21 6XS.

Palgrave Macmillan is the global academic imprint of the above
companies and has companies and representatives throughout the world.

Palgrave® and Macmillan® are registered trademarks in the United
States, the United Kingdom, Europe and other countries.

ISBN 978-1-349-38107-4     ISBN 978-0-230-10156-2 (eBook)
DOI 10.1057/9780230101562

Library of Congress Cataloging-in-Publication Data
Richardson, Jeanita W., 1957–
   The full-service community school movement : lessons from the
James Adams Community School / Jeanita W. Richardson.
     p. cm.
   Includes bibliographical references and index.

   1. James Adams Community School (Coatesville, Pa.)
   2. Community schools—United States—Case studies. 3. Children
with social disabilities—United States—Case studies. 4. Children of
minorities—Education—United States—Case studies.
   5. Segregation in education—United States—Case studies. I. Title.
LB2820.R53 2009
371.0309743′13—dc22                                                2009023746

Design by Integra Software Services

First edition: December 2009

10 9 8 7 6 5 4 3 2 1

Transferred to Digital Printing in 2013

So many people joined me on this journey but a few must be specifically named. Thank you to Diane Rofini and the Chester County Historical Society staff, Paul Johnson, Barry Gibney, Mrs. Betty Cuff, Vance "Carmin" Garnett, Aunt Teenie, David Anderson (Cuz), Aunt Angella, Wendy, and alumni and friends of James Adams. To Mom, Amanda, John, and all the invisible shoulders (like Grandmother White) I stand on who kept me going beyond my capacity, thank you. And...

To Grandma and Grandpa whose hands guided me beyond the veil to make sure their story was told in the spirit of CIHU. I love you and miss you more now than ever...

T. J. and Anita Anderson<br>
By: Jean Anderson White

Mom and Dad were an excellent team<br>
In our hearts they reign supreme<br>
Demonstrating love integrity and such<br>
Encouraging those whose lives they touched<br>
A special couple generating hope<br>
Leaders with courage helped us cope<br>
They taught us CIHU that the truth will set us free<br>
To be our very best, a better world we would see.

# Contents

Tables and Figures — viii

Introduction — 1

1 Full-Service Community Schools: Their Past, Their Present, and Their Future — 11

2 The Efficacy of a Community Approach — 37

3 Pursuing Excellence in a Segregated System: Coatesville, a Town of Paradoxes — 65

4 The Cornerstones: The Blurred Personal and Professional Lives of T. J. and Anita Anderson — 87

5 James Adams Community School: Manifesting the Vision — 111

6 In Their Own Words: Reflections from Alumni and Faculty — 139

7 Looking Backward to Plan Ahead: Lessons for the Community and Full-Service School Movement — 161

Index — 187

# Tables and Figures

**Tables**

| | | |
|---|---|---|
| 1.1 | Concentration of Students in Majority–Minority Public Schools | 15 |
| 1.2 | Community and Full-Service School Definitions | 18 |
| 2.1 | Leaders and Leadership | 52 |
| 3.1 | James Adams Principals | 80 |
| 4.1 | Community School Leadership Comparison | 106 |
| 5.1 | Partnership Comparison | 114 |
| 5.2 | Organizational Development and JACS | 123 |
| 5.3 | JACS Graduation Themes and Speakers | 129 |
| 6.1 | Benefits of a Highly Functioning Community School for Students and Communities | 157 |

**Figures**

| | | |
|---|---|---|
| 2.1 | From Risk to Resilience | 43 |
| 2.2 | Ecology of Human Development | 45 |
| 2.3 | Highly Effective Community School (HECS) Model | 50 |
| 4.1 | CIHU and the Community | 103 |
| 7.1 | Highly Effective Community School (HECS) Model Revisited | 169 |

# Introduction

> We have a powerful potential in our youth, and we must have the courage to change old ideas and practices so that we may direct their power towards good ends.
>
> – Mary McLeod Bethune (Bethune, 2000, p. 395)

Schools and the way they interact with children and families possess untapped potential. Unleashing the intellectual prowess of all of the nation's students requires a courage we have not yet collectively demonstrated. The bravery that educators need to exercise, as per Mary McLeod Bethune, is a willingness to break from traditional ideas and practices in order to embrace new and different strategies. For some children the standardized educational system in this country suffices. They are demonstrating acceptable degrees of success and in some cases are excelling. However, there are tens of thousands of other students who are languishing in classrooms across the country.

Public K – 12 schools are charged with meeting prescriptive accountability standards at a time when many children face dilemmas external to schools that directly and indirectly influence learning readiness and capacity. Tremendous resources ($440.3 billion in 2002–3) are being devoted to public schools and yet, schools alone do not command the resources needed to eliminate or mitigate the stressors to learning external to schools that impact learning readiness such as the specters of race, socioeconomic status, and health (Richardson, 2006).

In the wake of stringent standards and increasing numbers of "at-risk" youth failing to meet performance benchmarks, there is a growing acknowledgement that social networks are critical to academic success. In the late 1980s and early 1990s calls came from the academic literature to consider networks of organizations partnering with schools as an effective means of meeting the needs of children (Dryfoos, 1994; Kronick, 2005; Warren, 2005). Theorists such as Joy G. Dryfoos posited holistic interventions on behalf of children that by definition necessitated family and community inclusion (Dryfoos, 1994, 2002, 2003; Dryfoos & Maguire, 2002). The rationale was that schools are almost the last institution where all members of a community might express a vested interest. As scholarship and the practical application of holistic theories confirmed their value for children, (such as the groundbreaking Comer School Development Program at Yale (1968) founded by Dr. James P. Comer), in concert with the advocacy of Dryfoos, Blank, and others, the term "full-service community school" gained traction.

The Children's Aid Society defines community schools as a public school that combines the best educational practices with a wide range of vital in-house health and social services to ensure that children are physically, emotionally, and socially prepared to learn (The Children's Aid Society, 2007). More specifically, a full-service community school is a label for a facility that is open beyond the school day offering a broad array of health, educational, and social services that support children and families (Dryfoos, 2003). Given the presence of schools in virtually every community, the public dollars dedicated to their operation, the long-term economic ramifications of educational attainment for individuals and communities, and the likelihood that holistic interventions promote resilience, investments in community schools seem prudent.

The contemporary community school movement also reminds us of the old adage, "there is nothing new under the sun." Not only were community schools in existence at least as far back as the 1930s in West Virginia, but largely missing from the educational lexicon are accounts of sophisticated organizations functioning in the African American communities before enforced desegregation (Bertaux & Washington, 2005; Morris, 2004; Walker, 1996; Wuenstel, 2002).

One such school was the James Adams Community School, hereafter referred to as JACS. James Adams Community School was a segregated K – 9 school nestled in a small steel town, Coatesville, Pennsylvania. During their tenure (between 1943 and 1956), Thomas J. Anderson and his wife Anita envisioned and brought to fruition a fully functional operation where the building served as a school by day and a community center by night, incidentally the second time that T. J. Anderson was principal of the school. Virtually every facet proposed in the contemporary building blocks of full-service schools was in place at James Adams. For example, by day rigorous and creative curricula were implemented and every week night classes and activities were available to students, their parents, and members of the community at large, free of charge (e.g., tax preparation, literacy, and vocational classes).

JACS was not the only school of its kind. Yet, the mainstream history of U.S. public education is incomplete because of the dearth of records heralding school – community organizations that functioned in segregated cities and towns. When historical accounts of schools of the segregated era do exist, they have a tendency to focus on deficits and inequities associated with separate and unequal treatment as opposed to the strength these same schools demonstrated (Hilliard III, 2003; Morris, 2008; Perry, 2003). Thus, JACS adds to the literature base in that this school provides a vivid depiction of how even in the face of severely limited resources and segregation a community school can be created.

*The Full-Service Community School Movement: Lessons from the James Adams Community School* contributes in other dimensions to the educational literature through an articulation of T. J. and Anita Anderson's vision: how the community and faculty adopted the vision; what it meant in practical terms to matriculating students and their families; and an extraction of lessons applicable in the twenty-first century.

Given the accountability pressures on public schools and the research-based calls to more comprehensively support children in the twenty-first century, lessons to be learned from JACS have particular currency. Furthermore, as contemporary budgets are constrained at least in part by costs associated with standardized testing, T. J. and Anita Anderson's resourcefulness provides instruction on how to encourage community coalescence around the full-service concept with minimal resources. Their lasting impact is demonstrated more than 50 years after the dismantling of JACS in the lives of remaining students and faculty. Annually alumni meet near the site of the school where participants represent an impressive spectrum of professionals to include judges, actors, news broadcasters, educators, craftsmen, politicians, doctors, nurses, and entrepreneurs, each attesting to their JACS experiences as pivotal to their achievements. In the words of one alumnus, "there was something special about JACS. Mr. Anderson didn't take any stuff, but you knew he cared about you and you wanted to make him proud of you".

Much of African American tradition tends to be captured through oral accounts. While a valid and authentic source of historical events these orations can in some research quarters be discounted as unscientific. In this case, the oral tradition has been triangulated with a copious assemblage of documentation supporting verbal accounts. Rich data sources remain that verify not only what the Andersons accomplished, but also how they engaged the neighborhoods surrounding the school. For example, school activities were documented in local majority newspapers, personal documents, school archives, and can be triangulated through interviews with former students and faculty. Access to data and alumni has been enhanced by the relationship of the author to the Andersons, being one of their granddaughters.

Contributions of the proposed volume are also enhanced by the grounding of the JACS vision in the philosophy of John Dewey, acclaimed educational philosopher, who was on the faculty at Columbia University at the same time that T. J. Anderson matriculated as a graduate student.

Dewey advanced notions of community-centered schools as early as 1915 that were actively supported by Eleanor Roosevelt and implemented by Elsie Clapp (Warren, 2005; Wuenstel, 2002). Thus, from the perspective of the development of public educational paradigms, and preservation of inclusive historical accounts of public education in the United States, JACS has significance.

This manuscript is also timely because JACS served minority children with limited resources and was located in a high-need community. In a contemporary context, not only African American, but also American Indian and Hispanic American children are each most likely to live in high-need communities, attend low-resourced schools, and go to majority-minority schools (Children's Defense Fund, 2006; Freudenberg & Ruglis, 2007; The National Center for Education Statistics, 2008; Smith, 2003). Thus some of resourcefulness of the Andersons can be instructive to educational leadership serving similarly situated youth when recaptured in a systematic way.

## Methodology

Qualitative research by definition is inductive and analyzes a multiplicity of variables and their relationship interacting between and among one another in the context of a phenomenon, such as the creation of a community school (Yow, 2005). This investigation of JACS is constructed in the tradition of grounded theory and an oral history and as such, adhered to both sets of methodological standards. It is worth noting here the differences between an empirical methodology and method. Methodologies dictate ways of studying and thinking about a phenomenon. Methods, on the other hand, are the sets of procedures used to collect and analyze data (Strauss & Corbin, 1998).

Engaging in the research on JACS prompted a need to consider grounded theory, which was as an unanticipated undertaking. Grounded theory is a methodology that is derived from data and existing research resulting in a theoretical supposition that both emerges from the data and then is used to categorize and evaluate said data. This involved process yielded the Highly Effective Community School (HECS) model, described in detail in Chapter 2, which guides this investigation. Stated another way, the methodology of grounded theory was the lens through which the methods of oral historiography espoused by Yow (2005) were applied.

Oral historiographies are methods that can invoke a bit of a misnomer. Rigorous histories that are predominantly informed by interviews are also bolstered with documentation contextualizing the subject matter. Oral histories illuminate dimensions of life within a community. Objects, places, symbols, and their significance are contextualized in the stories of participants (Yow, 2005). That said oral histories are heavily influenced by narratives, which is in keeping with African American tradition. Historically, our stories have been passed down from generation to generation around dinner tables, at church picnics, in beauty parlors and barbershops. This is one of the reasons accounts of innovative and complex systems have been lost to contemporary audiences.

Qualitative methodologies also dictate a requirement that researchers disclose their relationships with participants and subject matter. These disclosures do not necessarily minimize the rigor of findings, particularly when the relationships between the researcher and participants allow for unprecedented access to data. Researcher bias is inherent in all forms of research, because it is the researcher who formulates questions and hypotheses and determines the types of data that will best answer research propositions. That said, qualitative researchers, particularly those who spend significant periods of time with participants and have preexisting relationships, must be careful to disclose the methodological standards adhered to.

Informants ranged in age from their late sixties to early nineties and as such interviews were deemed the most appropriate means to secure the above-noted data. Access to alumni and historical records was enhanced due to the relationship of the author to the Andersons. Interview techniques ranged from broad and open-ended questions to very structured ones. Limited memory of interviewees, imprecise articulation, and inaccurate assumptions by the researcher can inherently contain bias. Shortcomings are mitigated, however, by the richness of data available in the Chester County Historical Society Archives and the depth of understanding to be gained from those who attended or taught at JACS whose accounts concur with one another. For example, student and former faculty survey responses were compared to interview responses and discrepancies were resolved by virtue of triangulation.

Types of data being collected in this study that triangulated findings are institutional and individual documents, archival records, surveys, and interviews. Examples of documentation are correspondence, archived newspaper articles, school administrative documents, alumni interviews, and surveys. Personal files and pictures are also examples of archival documents. Standards for triangulation, rigor, and validity were applied to the study (Strauss & Corbin, 1998; Yow, 2005). While empirical standards inform the manuscript, you the reader have been spared research jargon in an effort to share the information most relevant to you.

A word here about the terminology used in the text. Over time the socially acceptable labels for citizens who by virtue of U.S. slavery practices came from Africa to this country have evolved. JACS functioned in a time when the terms Colored or Negro were most appropriate and since that time the terms have changed to Black and now African American. For purposes of clarity, all of the above noted identifiers will be used as synonyms. Similarly, depending upon the age of the publication, schools purposefully connected to communities have been called full-service schools, community schools, or full-service community schools. Each of these terms appears in this document as synonyms as well.

Chapters in this volume have been organized to build a case for the relevance of strategies applied at JACS for the contemporary full-service community school movement. It is noteworthy that the newly appointed Secretary of Education, Arne Duncan, has been supportive of this school model when serving as Superintendent of Chicago Public Schools (Duncan, 2003). Thus for those interested in a grounding in the best practices of community schools and why this alternative approach to education has great merit, this book is for you.

Chapter 1, "Full-Service Community Schools: Their Past, Their Present, and Their Future," provides a context for considering community schools by exploring their history in both White and Black communities. While structures vary, some of the earliest versions of this comprehensive model were documented in the 1930s grounded in the theoretical suppositions of John Dewey and operationalized by Dr. Elsie Clapp. African American communities had to create our own systems in the absence of the luxury of access and public affirmation. If we had not built or augmented our own hospitals, schools, and churches, they would not have existed or functioned well. Linking the past practices to contemporary demographic data is strategically employed to substantiate the notion that lessons from segregated community schools of the past are pertinent today.

Efficacy denotes the power or ability to produce desired results. When we consider the sordid successes and failures of the public educational system in the United States it becomes clear that shifts in practice that empower students and increase access to quality educational environments have not been initiated by legislators. Rather, advocacy at the classroom and at building levels have produced the most enduring positive results. Chapter 2, "The Efficacy of a Community Approach," provides an advocacy framework for community schools as well as advancing the HECS model as a way to not only examine institutions in retrospect, but also provide the foundation for contemporary effective and efficient full-service schools.

Comprehending how JACS came to be and the challenges involved with sustainability, one needs to place the school in its historical context. The next chapter, "Pursuing Excellence in a Segregated System: Coatesville, a Town of Paradoxes," recounts the era (1943–56) and the history of the town. Examples of schools similar to JACS provide evidence that community school innovations were not isolated to Coatesville and will perhaps prompt additional researchers to fill the gap in the educational lexicon of this period.

Leaders and their behaviors are the most significant predictors of organizational success. If we concede that organizations are associations organized for a specific purpose, schools are indeed organizations. Without exception, alumni, faculty, and public records confirm that JACS would not have been JACS without T. J. and Anita Anderson's leadership, and management acumen resulting in support from a broad cross sections of the city's businesses and public officials. The mantra of the school was "CIHU" ("Can I Help You"), which asserted community service as a responsibility and privilege of students and faculty. School day and evening programming as well as outreach initiatives are included to recreate the vitality that exuded from the school at the height of its operation. Information about the Andersons is detailed in Chapter 4, "The Cornerstones: The Blurred Personal and Professional Lives of T. J. and Anita Anderson," and synthesized into four aspects of leadership: structural, political, human relations, and symbolic attributes included as part of the HECS model.

Chapter 5, "James Adams Community School: Manifesting the Vision," provides insight into the organizational development and partners without whom the school could not have achieved full-service status. As was noted earlier, the typical school day was followed by a plethora of evening activities offered free of charge to the community. This section delves into specific partnership relationships to include with whom they were and how they functioned within the organization. Piecing together the impressive evening program offers great insight into the responsiveness of the building personnel to community priorities. That said burgeoning ranks of partners and programs could not have been sustained in the absence of evolving organizational capacity. Tenets espoused in the HECS model are the lens through which the partnerships and organization's development are analyzed.

It would be too easy to relegate this evaluation to an abstract theoretical exercise denying the fact that JACS is an example of ordinary people accomplishing extraordinary results. To bring Adams' vignettes to life, Chapter 6, "In Their Own Words: Reflections from Alumni and Faculty," tells the synthesized stories of parents, students, teachers, community members, and the Andersons drawn from multiple data sources. Additionally, the final section of the HECS model speaks about the benefits of community schools that can accrue to students and members of the community. After the compilation of stories, tangible and intangible benefits of being associated with JACS are revealed.

Finally, Chapter 7 extracts what we can learn from JACS to advance the contemporary full-service school movement. Admittedly there are strategies that are not applicable today because of the constraints imposed by federal, state, and local governance. However, in more cases than not, the tenacity of the Andersons and their partners are instructive in "Looking Backward to Plan Ahead: Lessons for the Community and Full-Service School Movement."

In the twenty-first century we have access to technology and resources unfathomable to the Andersons in the 1940s and 1950s. Yet, they managed through their leadership and engaged community partners to create a highly effective community school that would be enviable today. If they could accomplish so much with so little, surely we can do the same.

## References

Bertaux, N., & Washington, M. (2005). The "Colored Schools" of Cincinnati and African American community in nineteenth-century Cincinnati, 1849–1890. *The Journal of Negro Education, 74*(1), 43–52.

Bethune, M. M. (2000). Youth. In R. Newman (Ed.), *African American Quotations*, 395p. New York: Oryx Press.

The Children's Aid Society. (2007). *Community schools*. Retrieved October 25, 2007, from http://www.childrensaidsociety.org/communityschools

Children's Defense Fund. (2006). *Black child health fact sheet*. Retrieved September 24, 2007, from http://www.childrensdefense.org/site/PageServer?pagename=policy_ch_blackfactsheet

Dryfoos, J. G. (1994). Full-service schools: A revolution in health and social services for children, youth, and families. New York: Jossey-Bass.

Dryfoos, J. G. (2002). Full-service community schools: Creating new institutions. *Phi Delta Kappan, 83*(5), 393–400.

Dryfoos, J. G. (2003). A community school in action. *Reclaiming Children and Youth, 11*(4), 203–206.

Dryfoos, J. G., & Maguire, S. (2002). *Inside: Full-service community schools*. Thousand Oaks, CA: Corwin Press.

Duncan, A. (2003). *Chicago Public Schools Coalition for Community Schools: Press briefing*. Retrieved April 4, 2009, from http://www.communityschools.org/CCSDocuments/ADuncan.pdf

Freudenberg, N., & Ruglis, J. (2007). Reframing school dropout as a public health issue. *Preventing Chronic Disease, 4*(4), 1–11.

Hilliard, A., III, (2003). No mystery: Closing the achievement gap between Africans and excellence. In T. Perry, C. Steele, & A. G. Hilliard III (Eds.), *Young gifted and Black: Promoting high achievement among African-American students* (pp. 131–166). Boston: Beacon Press.

Kronick, R. F. (2005). *Full service community schools: Prevention of delinquency in students with mental illness and/or poverty.* Springfield, MA: Charles C. Thomas.

Morris, J. E. (2004). Can anything good come from Nazareth? Race, class, and African American schooling and community in the urban South and Midwest. *American Educational Research Journal, 41*(1), 69–112.

Morris, J. E. (2008). Research, ideology, and the Brown decision: Counternarratives to the historical and contemporary representation of Black schooling. *Teachers College Record, 110*(4), 713–732.

The National Center for Education Statistics. (2008). *The condition of education 2007* (Vol. NCES 2007064). Washington DC: U.S. Department of Education Office of Educational Research and Improvement.

Perry, T. (2003). Up from the parched earth: Toward a theory of African-American achievement. In T. Perry, C. Steele, & A. G. Hilliard III (Eds.), *Young gifted and Black: Promoting high achievement among African-American students* (pp. 1–108). Boston: Beacon Press.

Richardson, J. W. (2006). *Public K – 12 federal educational policy.* Battlecreek: The W.K. Kellogg Foundation.

Smith, J. (2003). *Education and public health.* Alexandria: Association for Supervision and Curriculum Development.

Strauss, A., & Corbin, J. (1998). *Basics of qualitative research: Techniques and procedures for developing grounded theory* (2nd Ed.). Thousand Oaks, CA: Sage.

Walker, V. S. (1996). Their highest potential: An African American school community in the segregated south. Chapel Hill, NC: The University of North Carolina Press.

Warren, M. R. (2005). Communities and schools: A new view of urban education reform. *Harvard Educational Review, 75*(2), 133–173.

Wuenstel, M. (2002). Participants in the Arthurdale community schools' experiment in progressive education from the years 1934–1938. *Chula Vista, 122*(4), 759–770.

Yow, V. R. (2005). *Recording oral history: A guide for the humanities and social sciences* (2nd ed.). Lanham: Rowman & Littlefield.

# 1

# Full-Service Community Schools: Their Past, Their Present, and Their Future

> The test of an advanced society is not in how many millionaires it can produce, but in how many law-abiding, hardworking, highly respected, and self-respecting loyal citizens it can produce.
> 
> – Franklin, J. H. (Franklin, 2005, p. 382)

Consensus is easily reached with respect to the notion that the challenges facing public schools are complex and multicausal. Empirical data go further to substantiate the fact that the one-size-fits-all educational approach has not alleviated the stubborn achievement gaps experienced by subsets of the nation's children. Viewed as one system, we in public education are earning failing grades in the test posed by Dr. John Hope Franklin when we consider the numbers of youngsters attending schools deemed substandard by virtue of a myriad of factors. That said few pundits have advanced strategies more grounded in historical success than the full-service community school model.

Educational institutions have been and continue to represent the gatekeepers of economic stability and quality of life for individuals. Rhetorically the public education system in the United States is in place to ensure each child is equipped to contribute positively to the national economy and civil society. In practice, however, the history of public education is one of emergent access for women, immigrants, African Americans, and American Indians. While K – 12 education in the public psyche seems to assume equality and equity, data and legislated enforcement of access would reveal otherwise. Consequently, queries about social equity often begin in schools and for good reason; gaps in academic outcomes are deeply rooted in the distribution of access to quality educational environments.

Most of America's children attend publicly funded K – 12 schools. In the academic year 2002–3, there were 14,465 public school districts and 95,615 schools. Of the 48.2 million students served, 4.1 million (8.5 percent) were limited in their English proficiency and 6.4 million (13.4 percent) were served by federally supported programs for the disabled. Nearly 52 percent of U.S. public school students attend suburban schools, 31 percent attend urban schools, and 17 percent are enrolled in schools considered rural (U.S. Department of Education National Center for Education Statistics, 2004). As per the U.S. Census Bureau, the years of majority White national school systems are quickly coming to an end. As of the 2006–7 school year 43.5 percent of pupils belong to racial/ethnic minorities (17.1 percent Black, 20.5 percent Hispanic\Latino, 4.7 percent Asian, and 1.2 percent American Indian) (Orfield, 2009). Of the children in attendance over 10 million (20 percent of the total school-aged population) speak a language other than English at home. Nearly 7 million students receive special education services, which raises the price tag for serving children in the public school system (The Annie E. Casey Foundation, 2007; Federal Interagency Forum on Child and Family Statistics, 2007; Planty et al., 2008).

What should give us pause are the conditions of the schools with concentrated populations of growing numbers of minority students, and the persistent academic outcome gaps. During the 2004–5 school year nationally only three-fourths of high school students graduated on time (in four years) (Planty et al., 2008). For African Americans and Latin Americans graduation rates are as low as 50 percent in some districts (America's Promise Alliance, 2008). Dropout rates or delayed graduation are measurable negative educational outcomes as are reading and mathematics test scores. Differences between Black and White fourth- and eighth-grade reading and mathematics scores range, while narrowing slightly over the last decade, have for the most part remained stagnant (Planty et al., 2008).

As per the National Center for Educational Statistics (NCES), African Americans, Hispanic/Latin Americans, and American Indians are more apt to attend schools labeled as "high-poverty schools" than their White or Asian counterparts. High-poverty schools are institutions at which more than 76 percent of pupils are eligible for free and reduced lunches (a federal government proxy for poverty) irrespective of locale (U.S. Department of Education – National Center for Education Statistics, 2008). An African American low-income child is more likely to attend a high-poverty school than a White youth whether they live in a city, town, suburb, or rural district. There is also a concentration of African American students in certain types of school districts. Urban centers are home to a disproportionate percentage of African American families, and as such schools are commonly majority-minority schools. Thus it is not uncommon for African Americans to represent in excess of 80 percent of an urban school population (Morris, 2008).

Schools are a critical contributor to growth and development; however, alone they cannot manage the complex, multidimensional inputs that positively or negatively impact youth (Jehl, 2007). While an in-depth analysis of poverty stressors and their attendant impact on schooling and academic achievement is beyond the purview of this book, given the significant numbers of children living at or below the U.S. poverty level, it is worth noting the relationships between economic deprivation and learning capacity in youngsters as a precursor to our discussion of community schools (Richardson, 2006). In 2002, more than 11 million children (16 percent of all youth under 18) lived in poor families. Children, because of their unique developmental processes and their inability to care for themselves, are most susceptible to the multiplied disadvantages associated with poverty. The longer that children live in poverty, the lower is their educational achievement and the slower are their general maturation processes—not because of innate deficits, but because of susceptibility to preventable diseases, toxic exposure, malnutrition, and other risk factors (Richardson, 2005, 2006).

Even short bouts of poverty, at critical junctures, can diminish cognitive capacity. Children faced with inadequate health care, food insecurity, and poor housing quality disproportionately experience concomitant threats to childhood educational potential because of the schools they attend. Rather than schools being a haven from the challenges of economic deprivation, they are instead magnifiers of the inadequacies of the very social system designed to mitigate dimensions of low status. For example, African American, Hispanic American, and American Indian youth are most likely to attend high-poverty schools (i.e., schools serving a predominantly poor population). These schools also tend to serve majority-minority populations and are poorly funded when compared to their suburban counterparts (U.S. Department of Education National Center for Education Statistics, 2004). Furthermore, cognitive and behavioral consequences of poor health can result in higher demand for costly interventions (e.g., special education services); unfortunately, high-poverty schools are the very institutions commanding the fewest resources (Richardson, 2005).

High-poverty schools had the highest teacher turnover rate in 2005–6 (21 percent), detracting from continuity and accrued pedagogical and content expertise. How big a problem is this; or are we considering a small segment of the school-aged population? Twenty-three percent of all public elementary and secondary students in 2005–6 attended high-poverty schools. However, even that disturbing number masks the concentration of minorities in said schools, because about 50 percent of all Latino and Black students attended high-poverty schools (U.S. Department of Education National Center for Education Statistics, 2006).

The sheer numbers of children dropping out of school and underperforming (a manifestation of risk exposure) are reason enough to consider a nontraditional approach to teaching and learning. Significant numbers of children are at risk of school failure because of factors outside the teacher – student exchange. For example, if children are experiencing social, emotional, health, or economic disadvantages, these influences infiltrate a youngster's ability to optimize their educational potential (Richardson, 2008). As a result, schools, if they are ever to "Leave No Child Behind," must cultivate partnerships with the communities they serve, in order to ensure students' holistic well-being (Kronick, 2005). To do so we need to critically assess hazards to student learning, as well as the multiplied disadvantage the quality of their schools imposes upon them. One threat reemerging in the twenty-first century is resegregation of many schools.

Theoretically the causes of resegregation are the many social, economic, and political factors that merge at the nexus of race and class. Most notably for Blacks and Hispanics, children who are also supported by economically disadvantaged families are relegated to neighborhood schools that happen to be substandard, at least in part because of public school funding formulas. As per Orfield (2009), segregation is so pronounced that in the 2005–6 school year economically disadvantaged Hispanics and Blacks were more isolated than at any other point of time since the death of Dr. Martin Luther King, Jr.. Viewing their schools in isolation with less prepared teachers, lower documented achievement, and lower graduation rates fails to paint a complete picture of life in many communities.

Since the 1970s to the present day, those who have commanded sufficient resources to move to more pleasant and safe housing arrangements have had the prerogative to choose where they live. Whites' and affluent Blacks' flight from cities across the United States has yielded a heavy concentration of homogeneous sections of cities, such a phenomenon being termed resegregation. Exit of the affluent has produced neighborhoods concentrated with low-income families because they could not afford to leave even if they wished to. The overrepresentation of African American and Latino/Hispanic American children in the ranks of the poor have by virtue of where they live provided a mirror image in their schools, yielding a system of *de facto* segregation. Additionally, segregation of students matters because 44 percent of public school students are non-White and that number is predicted to rise such that in a very few years, the school-aged population will be majority-minority (Orfield, 2009; U.S. Department of Education – National Center for Education Statistics, 2006). To put the numbers in context consider that in 2005, 23 percent of all public elementary and secondary schools enrolled over 75 percent students of color. Not only are African American, Hispanic American, and American Indian students more apt to attend schools where few if any Whites attend, they also are most likely to attend high-poverty schools. Majority-minority schools represent 45 percent of urban, 20 percent of suburban, 10 percent of town, and 7 percent of rural schools. This has spillover implications on funding and other resources for schools that the neediest children attend. For example, teacher turnover rates in high-poverty schools are higher than those in low-poverty schools. Rather predictably, achievement gaps among Whites when compared to Blacks, Latinos, and American Indians remain, particularly in the areas of reading and mathematics (U.S. Department of Education – National Center for Education Statistics, 2006).

As previously noted, the number of non-White students served by public schools is higher than it has ever been historically and is predicted to increase. Whereas in 1988–9 roughly 15 percent of students were Black and 11 percent were Latino/Hispanic, their proportions of the total school-aged population to 17 and 20 percent, respectively in 2005–6. Most students of color are concentrated in the West, South, and Northeast and in urban centers. Forty percent of Latino and 38 percent of Black students attend schools with a minority population that exceeds 90 percent, which demonstrates their concentration in schools of this type (see Table 1.1; Orfield, 2009).

It is not that segregated schools in and of themselves are causes of achievement gaps. Instead, it is more externally applied systematic and fiscal attributes of schools serving primarily African Americans and low-income populations that persistently contribute to disparities. Data

**Table 1.1** Concentration of Students in Majority–Minority Public Schools

| Race/ethnicity | % students in 90–100% minority schools 1988–9 | % students in 90–100% minority schools 2006–7 |
|---|---|---|
| White | 0.4 | 0.92 |
| Black | 33.5 | 38.5 |
| Latino/Hispanic | 33.4 | 40.0 |
| Asian | 12.9 | 16.2 |
| American Indian | 17.8 | 20.2 |

Source: Orfield, 2009.

confirm that schools serving significant populations of low-income students have also been shown to have a disproportionate number of members with residential instability, problems related to health and nutrition, high exposure to crime and gangs, and staffs weaker in preparation and performance than other schools (Orfield, 2009).

Developmental resources known to support learning are less likely to be afforded by Hispanic American and African American children. Based on one calculation Black and Latino children are half as likely to receive learning supports when compared to their White counterparts. No surprise then that performance gaps persist and students increasingly fail to graduate on time, if at all.

As per a recent report by the America's Promise Alliance (2008), 34 million children do not have access to adequate developmental resources. One in six 6-19-year-olds is overweight (at least in part due to the availability and cost of healthy food options), 34 percent of teens (12-17-year-olds) have no health insurance, 1 in 12 high schoolers has attempted suicide, and 1.6 million 12-19-year-olds have been victims of violent crime.

To add a layer onto external threats to optimal learning, there are educational policies that by virtue of their punitive accountability standards have slowed the adoption of innovative initiatives for students. For example, in the wake of policies such as No Child Left Behind (NCLB), teachers and administrators are forced to conform their practices to suit legislation that to date has done little to live up to its namesake's promise. NCLB is an ever-evolving policy whose predecessor was the Elementary and Secondary Education Act (ESEA), and it represents an iteration of long-standing federal public educational policy. Federal bodies extend their influence primarily through funding streams outlined by statute. Different from ESEA, NCLB adds corrective actions that carry punitive consequences for schools, districts, and educators. Whether intentional or not, poor academic performance has been placed largely at the feet of teachers rather than recognizing that learning and performance are the result of complex factors within and external to schools (Richardson, 2007, 2008). The reality is that contemporary accountability criteria have shifted classroom practice from a private exchange between teachers and students to fodder for front-page print.

We should remember that despite mammoth challenges young people still have dreams of working hard to improve their economic and academic lot (America's Promise Alliance, 2008). To foster youngsters' initiative and protect their potential administrators and teachers, need to abandon traditional approaches, particularly when they are not yielding desired results. As the aforementioned data allude to, since historic educational practices are not working, why not try something different?

Schools are one of the last institutions all communities have in common. Though difficult, many locales have proved that development and implementation of a full-service school is possible and yields positive results for children and their families. While more formalized as an educational model worthy of advocacy in the 1980s, community schools have existed in various forms in the United States as far back as 90 years ago. Even as schools forgo their separateness, meaningfully engaging communities is time consuming and hard work. By meaningful engagement I refer to an integration of community persons and organizations into the planning and implementation of programming. This is no small task given the public educational accountability standards captivating much attention.

The danger here is the divergent definitions of community engagement, which could range from having a meeting telling the community about programs to be offered, to soliciting community participation in priority setting and partnering. A useful definition of a school – community partnership is the purposeful connections between schools and external groups, such as businesses, social service agencies, and community-based groups that are forged with the goal of promoting children and youth's social, emotional, intellectual, economic, and physical growth (Sanders, 2001).

Before we go further it is important to acknowledge that substandard educational settings in no way imply that attending a high-poverty, majority-minority school is an academic death sentence. There have been numerous examples of youth emerging from such schools to excel academically and professionally. What is intended is an attempt to highlight the more intense needs of some students, whose educational paths are strewn with hurdles over which they have little or no control. In order to convert their educational environments and outcomes their schools need to become laboratories of tested approaches, one of which is the community-based model.

## Community Schools

Out of the divergent needs of families and educator's and a dissonance between school and community cultures, notions of a more integrated learning model have emerged. The need was grounded in an acknowledgement that fragmented approaches to teaching and learning, particularly for children experiencing circumstances known to create risk of economic, social, or educational failure, do not work well (Harris & Hoover, 2003). This was not however a contemporary epiphany. If one delves deeper into the paradigm under girding full-service community schools, one can find examples that significantly pre-date Dr. Dryfoos's work. Though much of the literature and advocacy advancing community schools gained momentum in the 1980s and 1990s, the foundational elements were articulated by famed educational philosopher John Dewey as early as 1915 (Bertaux & Washington, 2005a; Wuenstel, 2002).

A significant literature base has accrued over the last 20 years in support of community and full-service schools as models to more efficiently and effectively meet the needs of children. Most notably Joy Dryfoos has contributed volumes advocating for a model where a full range of social services are available to youth in the location they are most apt to spend significant time—their schools. Schools in her configuration are sites where a variety of services to include physical and mental health, enrichment activities, and tutoring are provided (Dryfoos, 1994, 2002, 2003; Dryfoos & Maguire, 2002).

**Table 1.2** Community and Full-Service School Definitions

| Community school definitions | Source |
| --- | --- |
| A school that is open most of the time; houses an array of supportive child and family health and social services provided through partnerships with community agencies; integrates quality classroom teaching with activities in extended hours; involves parents in significant ways; has a full-time coordinator; and serves as the hub of the community | Dryfoos, 2003, p. 203 |
| A full-service community school integrates the delivery of quality education with whatever health and social service are required in that community. These institutions draw on both school resources and outside community agencies that come into the schools and join forces to provide seamless programs. | Harris & Hoover, 2003, p. 207 |
| A community school is made with the people whose school it is. In the making teachers lead as fellow workers. As members themselves of the community, they are citizens as well as teachers, sharing common problems and interests. | Clapp, 1971, p. 89 (Student of John Dewey) |
| A community school is established to perform a job of total education.... An integrated day and night program caters to all elements of the population and seeks to bridge cultural gaps which are largely the result of unfulfilled group needs. The ultimate objective of the programming is to develop good American citizens who will make positive contributions to society. | Anderson, November, 1953 |

A community school does not exist as an island within a community; rather it is more of an oasis, where services and support are offered that benefit the broadly defined community. Table 1.2 advances several definitions of community and full-service schools.

Common across all definitions are several principles: the first is that the school adapts to the needs of the community and not vice versa. If the school is to be embraced by communities, it must be perceived as a sort of problem-solving center and be available beyond traditional school hours. Next the school staff and administration should demonstrate responsiveness to community-defined needs. Finally, the school integrates everyday, authentic values and problems into the curriculum and programming (Britton & Britton, 1970).

Also descriptions of community and full-service schools include their capacity to be conveners of vital resources for their distinctive neighborhoods. Because local challenges are so idiosyncratic, community schools take on district characteristics. For example, consider a neighborhood served by a school that has many undereducated adults whose low-wage jobs fail to provide basic health benefits for children. A vital function of that community school would be to offer GED classes, job training, and health services after traditional school hours. The school has the potential of functioning as an effective socializing agency as per Dewey (2001/1915).

Configurations of community schools can vary widely and can include but not be limited to health and social service agencies, school-based health-care clinics, family support groups, youth development organizations, government-sponsored programs, vocational and educational classes targeting adults, student tutoring and mentoring, and basic need services such as clothes closets and food pantries. Fundamentally, the myriad activities and services are designed to support children and their families at the point of their need with a goal of optimizing economic, academic, and social potential. Full-service schools work because the organizations are respected entities and, as a result, have the cache needed to broach certain partners that might otherwise not invest in disadvantaged neighborhoods. Perhaps most important is the creation of a welcoming atmosphere for community residents through cooperative identification of challenges and the empowerment of families.

In theory, one would be hard-pressed to find detractors of community schools. Support wanes in the details around how practical the partnerships would be and the costs associated with implementation (Keith, 1999). Contemporary calls for community schools suggest the inefficiencies associated with teaching children while attempting to ignore other dimensions of their lives external to the school is unnecessarily expensive. In support of the model advancing a more holistic interventions for children and youth, some of the reported benefits of community schools have been:

- improved readiness to learn;
- extended learning opportunities to children and adults;
- increased parental involvement;
- provision of opportunities to integrate community services;
- enhanced safety within communities;
- provision for basic needs of children; and
- mitigation of school failure (dropouts, academic performance) (Blank, Melaville, & Shah, 2003; Dryfoos & Maguire, 2002; Kronick, 2005; Warren, 2005).

Evaluations of benefits that accrue to youth, particularly in urban settings, have demonstrated that despite the presence of factors indicative of educational, economic, and health risk, the presence of a community school does much to insulate youngsters from distress. One key admission to reversing negative predictions for children has been the acknowledgement that children do not learn well if they are hungry or sick, or live in inadequate housing. In addition, teachers ill equipped by virtue of ignorance of their students' cultures and lives are not poised to promote success in the classroom (Warren, 2005).

John Dewey's 1915 conceptions of democratic, community-centered educational settings are informative as we consider contemporary appropriations of the model (Warren, 2005). Dewey posited that schools function best when they advance community purposes. By making health, recreation, and occupational advancement central to a school's mission, the entire community gains (Clapp, 1971). Community-centered schools as per Dewey follow more of a communitarian as opposed to individualistic paradigm. Individualistic education concerns itself only with teacher – pupil interactions, specifically around learning a curriculum. A hidden danger of the mere absorption of facts, in his opinion, is its tendency to lure children into selfishness and competitiveness (Dewey, 2001/1915).

Conversely a communitarian paradigm supports the notion that learning is contextually relevant and contributes to positive social engagement. As such, instead of learning remaining abstract, applying course content provides opportunities to see how as an individual one can contribute to the greater good. Stated a different way, the school is an embryonic society inseparable from the well-being of its community. Dewey went on to say that creating an isolated educational experience is counterproductive as it pertains to helping children draw inferences to the interconnectedness between ideas and life. He posited that the only way to unite learning and life was to model the integration in schools, that is, "we must look at it [schools] as part of the larger whole of social life" (p. 45). He believed so much in a community-centered approach to schooling, that he went as far as to say that failure to provide students an opportunity to apply learning to daily living was wasteful (Dewey, 2001/1915).

Fundamentally, schools were in Dewey's eyes communities within communities, which in the absence of relating to the daily lives of students were misappropriated. Seamless integration of curricula to the lives of children would nurture and demonstrate that learning was not an isolated activity. He also cautioned that schools could not replicate models identically. Rather, the planning of a community school took into account the feasibility of implementation; this should be informed by other such schools and practices and then tailored to the particular demographic, social, and economic needs of the neighborhoods served (Dewey, 2001/1915).

While Dewey spent much time advocating for the aforementioned ideas it was his daughter and one of his students, Elsie Clapp, who implemented the tenets of his theory. In the 1910s and 1920s, Evelyn Dewey and Marie Harvey converted the Porter School, in rural Missouri, into a community school. In addition to correlating subjects to seasons and the life experiences of students, learning centers were created to foster curiosity. Transportation was an issue for many students and in response a horse-drawn carriage became a first-generation school bus in 1912. Adults were welcomed into Porter with reading clubs, a farm women's club, nondenominational church services on weekends, and various agricultural clubs. Bands, which boasted student and adult performers, performed locally and even at the state fair (Ediger, 2004).

Elsie Clapp, developed community schools in Kentucky in 1928–9 and in West Virginia during 1934–8. At the time, her initiatives coincided with the Roosevelt-era attention to impoverished communities and community empowerment. President Roosevelt and in particular Mrs. Eleanor Roosevelt visited the schools, and provided political and in some cases fiscal support for the model (Clapp, 1971; Wuenstel, 2002).

Clapp's (1971) leadership and implementation of the Arthurdale School in rural West Virginia were recorded in a book entitled *Community Schools in Action*. Influencing success was the role of teachers because they did not examine and evaluate the community, they were part of it. Using Dr. Clapp's terminology, "a public school is a school of the people of that district" (p. 61). Problems central to needs were contextualized after getting to know families and determining community assets. In response what surfaced were daily health and cleanliness checks, dental service, chronic disease management, well-baby clinics, immunizations, plays, concerts, community socials, and adult education, all as part of programming. As a specific example, a pressing health issue at the time was tuberculosis, and as such screenings and vaccinations were offered in the school for students and their families (Clapp, 1971).

Parents were often found in the school working side-by-side teachers. Arthurdale School operationalized the notion that there should be no distinction between when and where school and life outside the school begin and end. Perhaps teachers at the time articulated the climate best. Though sometimes it seemed to educators they were being commissioned to make bricks with straw, Katherine Kimble said, "we saw our work as building an environment for health and opportunity" (Clapp, 1971, p. 336).

Community schools as a politically supported intervention did not reappear in the public consciousness until the late 1980s and 1990s (Warren, 2005). Many organizations and foundations have supported their creation and that of neighborhood-linked schools by virtue of finances, advocacy, or research. Some contributors are noted below:

- Coalition for Community Schools
- National Center for Community Education
- C.S. Mott Foundation
- Jane Addams
- University of Pennsylvania's – Beacons
- United Way Bridges for Success
- Coalition for Community Schools
- Fordham Center for School Community Partnerships
- Family Support America
- National Collaboration for Youth
- George Soro's Open Society Institute

Given the increasing popularity of community schools as a model for public education delivery it is important to consider the contemporary benefits associated with implementation.

Consider three elementary schools in modern Knoxville (Tennessee), which employed the full-service approach gradually at first by offering after-school classes. Later they added services informed by surveyed parents that included but were not limited to health clinics, GED classes, computer training, financial management, and interviewing skills. The results of systematic community engagement translated into increased student attendance, fewer students being unsupervised in the evening, fewer students exhibiting disruptive behaviors (during and after school), and increased time on-task during the school day. Measurable outcomes were attributed to customized classes and activities and ongoing engagement with parents and members of the broader community. Most notably, children's attendance seemed largely improved because youngsters were feeling successful and excited to see their mentors, and were healthier (a school-health clinic was opened in the school, staffed by local medical students and a physician, who donated services *pro bono*) (Kronick, 2005). Viewed alone, the success of the Knoxville example might seem happenstance. However, many such examples exist and at varying levels are demonstrating the value of this paradigm shift in public education. In 2000, the George Washington Community School, Indianapolis, Indiana, funded by the Anne E. Casey Foundation, served 900 seventh- to twelfth-grade students. Approximately 90 percent of the student body qualified for free or reduced lunch (a U.S. Department of Education marker for poverty). Forty-nine partners worked to provide services and resources for students and their families. Rooted in the prevalence of risk factors, students in this school would predictably perform poorly on tests. However, in the first graduating class in 2006, 100 percent of students passed the state's Graduation Qualifying Exam in mathematics and 90 percent passed the language arts portion of the same test. Not only did students demonstrate great resilience, they outperformed all the traditional high schools in Indianapolis (Jehl, 2007).

A study commissioned by the Coalition for Community Schools of 20 community schools revealed that the model consistently contributed to improved student learning by addressing the physical, social, and emotional needs of students. Fifteen of the 20 schools reported an increase in grades and standardized test scores; improved levels of family engagement; increased attendance; and improved broader community engagement with the school—thus adding a level of historically absent vitality (Blank, Melaville, Shah, 2003).

An elementary school in Green Bay (Wisconsin) reported an increase in third-grader reader proficiency scores to 61 percent up from 40 percent several years earlier. Another elementary school in Chicago, despite over 96 percent student poverty rate, showed improved reading scores, which were greater than the citywide average. An East Hartford (Connecticut) high school reduced dropout rates to 2 percent from 22 percent over a six-year period. A Carson (California) high school noticed a 50 percent reduction in suspensions, which are results akin to a Lincoln (Nebraska) elementary school that saw an 80 percent decrease in disruptive student referrals. As a final example, an elementary school in Indianapolis, because of collaborations with the health community, had 100 percent of their kindergarteners and fifth-graders receiving the full complement of immunizations (Blank et al., 2003).

Benefits of community schools are both intuitive and grounded in research. Children whose health, social, economic, emotional, and social needs are met are much more apt to attend school regularly and perform in accordance with their innate intellect. The converse is also true. If needs are unmet, well-being and educational acumen are compromised. It is a false dichotomy for educational leaders to choose between meeting contemporary accountability standards (teaching to the test) and forming a community school, particularly when the makeup of the student body indicates familial and community problems that can negatively impact learning.

Freedom granted to principals by their respective superintendents has a great deal to do with their ability to create a community full-service school. In addition, No Child Left Behind (NCLB) and the Individuals with Disabilities Act (IDEA) have each advanced stringent accountability standards that preclude in some cases and diminish in other cases the likelihood that a district will "experiment" on a community school. Nevertheless, given the demonstrated success of this educational model and the children, they are apt to aid disparities in educational settings and concentrated communities of risk beg us to consider ways to intervene.

Whether by design or need, full-service schools tend to be located in disadvantaged neighborhood (Dryfoos, Quinn, & Barkin, 2005). By disadvantage I refer to economic, social, or ethnicity- or race-based factors, which, as per the prevailing literature, are predictive of educational underachievement. Underachievement for our purposes is a failure to attain educational proficiencies one's unencumbered intellect would predict. Poverty and/or race and the multidimensional ways in which they impact children is the "disadvantage" most often referred to in conversations of learning potential.

There appears to be no better time to create community schools for children because those in greatest need of assistance are the ones attending traditionally organized schools that are failing to meet established standards. Students populating U.S. public education classrooms are more diverse than any time in our history, yet stubborn inequities persist. Despite the 55 years post-*Brown*, the state of equity in education for African Americans and other racial and ethnic minorities remains a dream unfulfilled. Of particular relevance to this book is the advocacy that prompted the slow methodical movement toward equitable opportunity for Black children, forcing communities to create institutions meeting their needs rather than waiting for government to make good on its promised interventions. Now that the disparities in public education and the potential of community schools has been established it is time to turn our attention to the characteristics of a well-functioning full-service community school.

### Hallmarks of a Well-Functioning Community School

Benchmarks of a successful full-service community school and the ways one demonstrates effectiveness articulated by the Coalition for

Community Schools are in accordance with the recommendations of other researchers in the field (Blank et al., 2003; Dryfoos & Maguire, 2002; Kronick, 2005). Examples of conditions for learning in a community school guiding this examination are the following:

- The school has a core instructional program with qualified teachers.
- Students are motivated and engaged in learning both in school and in community settings after school.
- The basic physical, mental, and emotional health needs of young people and their families are recognized and addressed.
- Community engagement together with school efforts promote a school climate that is safe, supportive, and respectful and that connects students to a broader learning community.

Characteristics of a good community school are strong leadership, collaborative partnerships, and organizational capacity, which impact both tangible and intangible student outcomes. The Coalition for Community Schools articulated five conditions for learning that organizations/schools need to address in order to adhere to the model. First, the school must have core instructional programs implemented by qualified teachers. Curricula also need to be based on standards with clear expectations for student proficiencies. Second, programming should be designed to keep students motivated and engaged in learning both within traditional school hours and in community settings. Third, services should address the basic physical, mental, and emotional health of youngsters and their families. Fourth, an atmosphere of mutual respect and effective collaboration between and among parents, community members, and school staff needs to be in place. Finally, community engagement in concert with school-based initiatives promote supportive, nurturing, and practical programming informed by idiosyncratic community needs (Blank et al., 2003).

Furthermore Blank et al. (2003) have advanced the characteristics of a stellar community school leader, which include:

- motivating a vision;
- connecting activities to the learning experiences;
- developing community partnerships; and
- fostering a well-functioning organization with requisite financing.

Planning a community school requires a holistic approach, defined as consideration of the full scope of health, family, emotional, and social needs that potentially influence learning. In the words of Elsie Clapp, if a community school is to align with its theoretical underpinnings, the answer to the question, "where does school end and life outside begin?" is answered with "there is no distinction between them" (Clapp, 1971, p. 89). In order to accomplish this feat, educators need to relinquish at least in part notions of separateness and professional turf by soliciting community partnerships that in the long run better serve the interests of children (Harris & Hoover, 2003).

Keeping schools open beyond traditional school hours can be a challenge. For example, there are insurance implications and utility and security costs associated with extending the school day. However, these hurdles must be surmounted if the model is to maintain its integrity. Parents and caregivers, particularly in middle- to low-income neighborhoods, tend to work hours that do not lend themselves to easy participation during the traditional day. Taking that into account, volunteers between nine and four may need to come from other groups. Even though initial visioning and implementation is labor intensive, the benefits of extended programming seem to outweigh the costs for students most sensitive to factors detracting from their educational experience.

Leadership styles have been categorized as transformational or transactional by some researchers. Transformational leaders inspire "followers" to adopt a common value system, work ethic, and organizational goals as congruent with their own sensibilities and inner moral compass. Transactional leaders apply companions of rewards and punishments to achieve organizational goals (Goodwin, Wofford, & Whittington, 2001; Popper, 2000). The first manner elicits a level of commitment that is internally generated and in practice does not require the principal to constantly cajole teachers and partners. The second requires the principal to constantly gauge the desired outcomes and mete out affirming or discouraging feedback. A sustainable community school requires at least in the planning and implementation stages a transformational leader. A similar leadership label would be charisma. A charismatic leader in practice would devote their efforts toward service, identifying points of mutual consensus and benefit with "followers," maintaining two-way communication as opposed to hierarchical procedures and relying upon the mores of "followers" to guide implementation of collaborative work (Popper, 2000).

**African American Public Education and Community Schools**

Largely missing from the literature espousing the value of full-service schools is an acknowledgement of the stellar examples of community – school partnerships prevalent in the African American community before and after the dismantling of legalized segregation. One contemporary example of how even in the face of poverty, residential segregation, and a host of textbook hindrances to academic excellence, outstanding outcomes were still achieved is the Fairmont Elementary School in St. Louis, Missouri. The staff and faculty had developed strong connections to the neighborhood's African American families and students. It not only had one of the highest attendance rates in the city, but its students also regularly outperformed other students (including those of the magnet schools) on standardized examinations. A similar study was conducted at Lincoln Elementary School in Atlanta (Georgia) with parallel findings (Morris, 2004). Empirical examinations provide hints that there is much to be learned by investigating segregated antecedents of contemporary community schools. Thus we begin our journey backward to James Adams Community School.

It may seem odd that a town in the proverbial North would impose segregated schooling on Black children. However, whether by formal or informal policy, segregated schools were the norm for African American children until after the *Brown v. Board of Education* ruling. Subsequently, the attention to the inequitable distribution of resources to support the schooling of Black children and the related prevalence of poor facilities and materials when compared to schools servicing Whites has been well documented. Thus it is not the memory of inequality that is incorrect, but rather **it is** the evidence affirming that vantage point alone **that** is incomplete (Walker, 1996).

Examples of stellar schools of the James Adams Community School (JACS) ilk have been documented in the work of Vanessa Siddle Walker (1996), Jerome Morris (2004, 2008), James Anderson (1988), and others. Community schools of Cincinnati (Ohio), Little Rock (Arkansas), Caswell (North Carolina), and Tuscumbia (Alabama), all had schools pre-*Brown* that met at least in part the contemporary definition of a community school (Bertaux & Washington, 2005b; Morris, 2008). Participants in these educational ventures were every bit as resourceful and committed as those affiliated with JACS. This volume attends to the principles from a historical context most useful for the contemporary community school movement.

It is germane at this juncture to clarify distinctions between the terms desegregation and integration because of their prolific use in the literature. Desegregation speaks to provisions articulated in law or practice that eliminate the isolation of members of a particular group into separate functional units. Integration, the goal of the National Association for the Advancement of Colored People (NAACP) litigation team, requires the incorporation of individuals and groups as equals into society. There has been much discussion whether desegregation has ever taken place and integration remains an ideal. This distinction is important because the goal of Blacks in the examined period, 1943–56, was specifically integration.

*Brown* was actually a collection of five cases from Kansas, South Carolina, Virginia, Delaware, and the District of Columbia. Though distinctions existed among the cases, the common thread was the argument that *Plessy*'s "separate but equal" ruling violated the equal protection clause of the Fourteenth Amendment. The significance of hearing the cases collectively under the umbrella of *Brown v. Board of Education* signaled the court's perception that public school segregation represented an issue of national scope and importance as opposed to a regional dilemma (Richardson & Harris III, 2004).

Yet, *Brown* for all its accolades also yielded unintended consequences from which the African American community has never recovered. Given the unfair analytical advantage of hindsight, at least in the case of lower schools (pre-K – 12), *Brown* implied that Black schools were inherently inferior to White schools (Wilkerson, 1979). Perhaps in part because of the premise of the case, unanticipated consequences of the favorable ruling have decimated functional social networks. We have witnessed a loss of Black educators; thwarted learning opportunities for Black children because of dominant cultural practices; decreased control over the socialization of children leaving them less prepared for the inequities they face; and diminished Black parental involvement in schools (Brown & Harris III, 2004; Hilliard III, 2003; Morris, 2008).

*Brown*, while incredibly influential, was a product of its time. Integration held the potential of forcing the nation to make good on its promise of equal protection and appeared to be a way to ease the social burdens associated with separate school systems by equipping youth for the future through access to equitable educational resources. Justices and the plaintiffs' litigation team believed that school segregation purely by race relegated African American youth to the status of second-class citizenship (Richardson & Harris III, 2004).

In retrospect integration was a risky pursuit, particularly as it related to the psychic security of children. African American children lost innumerable opportunities for development and mentorship post-*Brown* because they came to be judged by standards conceived and implemented outside their cultural and ethnic communities (Brown & Harris III, 2004). Youth were no longer taught by persons who looked like them, or individuals who had successfully negotiated the Eurocentric educational gauntlet as an outsider. They were and are taught in most cases by Whites. Verification of the racial distribution of teachers is easily confirmed; of the total number of K – 12 teachers in the United States only 14 percent are teachers of color (Irvin, 2004; National Education Association, 2003).

Even in light of judicial decisions such as *Brown*, widespread progress has not been made in eradicating the deep-seated presumptions undergirding racism. There was and continues to be irreconcilable tension between professed national beliefs and actions (or inactions) that perpetuate inequities determined by one's race (Richardson & Harris III, 2004). Despite its good intentions a spillover impact of desegregation was the fracturing of a complex community network of organizations and resources (Morris, 2004). Churches, schools, and Black professionals and shopkeepers all worked and lived together. They needed one another to insulate themselves and their children from negative stereotypes and Jim Crow practices outside their respective neighborhoods.

From more than one vantage point, the systematic deconstruction of the supportive web such as was present in some communities is in need of replacement and the community school movement is at its root a return to what was the historical African American tradition. This is precisely why an in-depth examination of a segregated community school is so relevant in the twenty-first century. Another reason lessons from the segregated era of education are pertinent can be found in the literal changing complexion of school populations. The *Brown* decision espoused the inherent unequal nature of segregated schools, yet as a function of housing availability, a reversal of Supreme Court support for enforced integration, and other social and economic factors, many schools in the first decade of the twenty-first century are markedly segregated and becoming more so as time passes (Orfield, 2009).

While government has a responsibility to care for all citizens, waiting for equity and equality has not been a timely means of ensuring support for many communities. This has particularly been the case for African Americans who by virtue of slavery and formal legislation have experienced a unique form of disenfranchisement. This is not to say that other groups such as American Indians, Latin Americans and other immigrant groups have not been discriminated against. Each has their own story to tell. The focus of this book is the educational innovation rooted in the African American community. Change, as well as supporting and preparing our children, has historically been instigated at the community level (Morris, 2004). This is also what makes lessons learned from JACS relevant to a contemporary full-service movement.

Segregation enforced in both the North and South spurred the creation of schools that served as an oasis of the community in the pre-Brown era. By virtue of active and sustained collaborations, the Black community did not wait for opportunities to come to them, but they created trajectories for themselves. One such example was JACS. While headed by other principals, it was the arrival of Thomas Jefferson and Anita Anderson that propelled this small K – 8 school to a full-service institution in every sense.

JACS was a segregated K – 9 school in a small southeastern Pennsylvanian steel town. With far fewer resources than are available today, T. J. Anderson and his wife Anita envisioned and brought to fruition a fully functional operation where the building served as a school by day and a community center by night between 1943 and 1956, after which the Supreme Court decision forced the gradual closure of the school.

Schools like JACS organized themselves as "as-if" havens. Out of oppressive Jim Crow practices, separate and unequal funding and facilities, as well as prevailing notions of inferiority, they functioned as though they were free people. As liberated persons they deemed themselves capable of controlling their destinies and able to become anything they fathomed. The total school – community ethos was built around this theme and was passed to students through curriculum, recreational activities, civic engagement, and creative venue (Perry, 2003). This approach embodies the whole idea of schools functioning as a protective factor against risk, which will be discussed in depth in Chapter 2. For now, schools as protective factors distinguished themselves with high expectations, exposure to Eurocentric and Afrocentric canons, and a plethora of opportunities to practice and identify special competencies. Students, as a result of the targeted mentoring, functioned "as-if" society was egalitarian. They were emboldened by self-esteem and equipped to combat inequality wherever they encountered it.

Another way to conceptualize this "as-if" schema is to think of Black community schools as institutions advancing a systematic counter-narrative relative to their students (Morris, 2008; Perry, 2003). The narrative is not much different than some of those espoused even today. African American youth have the dubious distinction of being part of a group whose intellectual prowess is perpetually considered in literature as well as debated in ideologies and policies. The counter-narrative was operationalized as teachers and principals intentionally opposed the ideology that their children were intellectually inferior and set out to prove their belief to be true (Perry, 2003). Failure to compete on par with Whites in their estimation had less to do with innate markers and more to do with exposure, curricular rigor, and supportive environments.

Admittedly, there are differences between the legal and environmental landscape of schooling between the days of JACS and now. For example, curricular development is constrained by compliance edicts tied to legislation such as the 2001 NCLB with its heavy reliance on standardized test scores to demonstrate content competence. Segregation is no longer *de jure* (by law) but a function of *de facto* (by practice) factors, and neighborhoods are not as cohesive as they once were. By that I mean that when Blacks were permitted to live where they pleased, the by-product became a dispersal of the professionals (white-collar and blue-collar workers) who once resided together. Everyone's children used to attend school together, thus providing incentive to make sure the environment was supportive of educational achievement regardless of class. It can be argued in a contemporary context that the residential mobility that middle income and the post-*Brown* era created disassembled vibrant layered neighborhoods. The subsequent outward migration of Black professionals to suburbs in search of better housing and school reputations has conversely concentrated the students in greatest need of role models to a second-class educational system.

However, the similarities between the time of JACS and today outweigh the differences. As the data verify, there are schools in the United States, particularly in urban centers, that host homogeneously Black or Latino populations with a high incidence of poverty. Given the funding schema of public education and the likelihood that schools in poor neighborhoods are also "poor" fiscally, they are as disadvantaged as legally segregated schools were. Data can also lead us to the conclusion that at the convergence of majority-minority schools and poverty, attendant educational risks are predictable. Yet, despite these dire predictions, both in a historical and in a contemporary context, there are iconic exceptions that confound negative outcomes, particularly when the community school model is applied.

Given contemporary standards of community school effectiveness, what can JACS teach us today? Though the populations of children and youth attending public schools in the United States are more heterogeneous than at that time in history, the segregation, particularly of African American and Hispanic American pupils, is more common than one might think. Adding more data to support this premise, 40 percent of all Blacks in the United States live in 11 central cities. This population concentration in addition to the higher rates of child poverty in cities (31 percent as opposed to 13 percent in suburban neighborhoods) provide hints regarding the populations served by urban educators (Riehl, 2000).

Funding disparities between affluent and economically distressed communities are not markedly different than was the case in the 1940s and 1950s. Enclaves of Black communities then and now are sequestered in neighborhoods where the local tax base (because of the location and type of housing) negatively impacts the dollars available to support education. Serving high-need student bodies without resources comparable to at least their suburban counterparts make the resourcefulness of the JACS community informative to us today.

African American teachers have always been central to the dismantling of educational and social inequities grounded in systematic governmental disenfranchisement of youth, if only by virtue of educating students beyond the capabilities majority culture had ascribed to them. Educators it can be argued are some of the most eloquent advocates for children and families today, because they know firsthand the potential thwarted because of antiquated practices, inequitable funding, and imposed constraints in the name of accountability that limit creative solutions that would benefit students. Educational attainment as a tool for social mobility was and continues to be the rhetorical value of schooling and inclusion of broader segments of society, which cannot be attained without a purposeful inclusion of teachers in the change process.

Lest we forget, educational attainment is also heavily influenced by children themselves, who come to school with their own issues that can determine their learning readiness. The tangled web of politics, teacher preparation and biases, and childhood challenges such as poverty are just a few strands that impact the ability of educators to ignite the spark of learning. The inherent value of shifting the delivery focus to a community and full-service approach is that it honors the assets of a community and capitalizes on practices that have historically been documented as effective, such as was the case for students at JACS.

## References

America's Promise Alliance. (2008). *Every child every promise: Turning failure into action*. Retrieved January 31, 2009, from http://www.americaspromise.org/uploadedFiles/AmericasPromiseAlliance/Every_Child_Every_Promise/ECEP_Reports_-_JPEG/ECEP%20-%20Full%20Report.pdf

Anderson, T. J. (November, 1953). *Coatesville city school district minutes*. Coatesville: Chester County Historical Society Archives: Coatesville City School District.

The Annie E. Casey Foundation. (2007). *2007 Kids count data book*. Baltimore: Annie E. Casey Foundation.

Bertaux, N., & Washington, M. (2005a). The "Colored Schools" of Cincinnati and African American community in nineteenth-century Cincinnati, 1849-1890. *The Journal of Negro Education, 74*(1), 43–52.

Blank, M. J., Melaville, A., & Shah, B. P. (2003). *Making the difference: Research and practice in community schools*. Washington, DC: Coalition for Community Schools, Institute for Educational Leadership.

Britton, J. O., & Britton, J. H. (1970). Schools serving the total family and community. *The Family Coordinator, 19*(4), 308–316.

Brown, F., & Harris, J. J., III (2004). Introduction: Did Brown fulfill the dreams of the second reconstruction congress? *Education and Urban Society, 36*(3), 239–244.

Clapp, E. R. (1971). *Community schools in action.* New York: Arno Press & The New York Times.
Dewey, J. (2001/1915). *The school and society & the child and the curriculum.* Mineola: Dover Publications.
Dryfoos, J. G. (1994). *Full-service schools: A revolution in health and social services for children, youth, and families.* New York: Jossey-Bass.
Dryfoos, J. G. (2002). Full-service community schools: Creating new institutions. *Phi Delta Kappan, 83*(5), 393–400.
Dryfoos, J. G. (2003). A community school in action. *Reclaiming Children and Youth, 11*(4), 203–206.
Dryfoos, J. G., & Maguire, S. (2002). *Inside: Full-service community schools.* Thousand Oaks, CA: Corwin Press.
Dryfoos, J. G., Quinn, J., & Barkin, C. (2005). *Community schools in action: Lessons from a decade of practice.* Oxford: Oxford University Press.
Ediger, M. (2004). Evaluation of the community school concept. *College Student Journal, 38*(1), 3–9.
Federal Interagency Forum on Child and Family Statistics. (2007). *America's Children: Key National Indicators of Well-Being 2007.* In Federal Interagency Forum on Child and Family Statistics (Ed.). Washington, DC: U.S. Government Printing Office.
Franklin, J. H. (2005). *Mirror to America.* New York: Farrar, Straus and Giroux.
Goodwin, V. L., Wofford, J. C., & Whittington, J. L. (2001). A theoretical and empirical extension to the transformational leadership construct. *Journal of Organizational Behavior, 22*, 759–774.
Harris, M. M., & Hoover, J. H. (2003). Overcoming adversity through community schools. *Reclaiming Children and Youth, 11*(4), 206–211.
Hilliard A. G., III. (2003). No mystery: Closing the achievement gap between Africans and excellence. In T. Perry, C. Steele & A. G. Hilliard, III (Eds.), *Young gifted and Black: Promoting high achievement among African-American students* (pp. 131–166). Boston: Beacon Press.
Irvine, J. J. (2004). Gap shrinkers. *NEA Today, 22*(8), 30.
Jehl, J. (2007). *Connecting schools, families & communities.* Baltimore: Annie E. Casey Foundation.
Keith, N. Z. (1999). Whose community schools? new discourses, old patterns. *Theory into Practice, 38*(4), 225–234.
Kronick, R. F. (2005). *Full service community schools: Prevention of delinquency in students with mental illness and/or poverty.* Springfield: Charles C. Thomas.
Morris, J. E. (2004). Can anything good come from Nazareth? Race, class, and African American schooling and community in the urban south and midwest. *American Educational Research Journal, 41*(1), 69–112.
Morris, J. E. (2008). Research, ideology, and the Brown decision: Counter-narratives to the historical and contemporary representation of Black schooling. *Teachers College Record, 110*(4), 713–732.

National Education Association. (2003). *Rankings and estimates: A report of school statistics*. Washington, DC: Author.

Orfield, G. (2009). *Reviving the goal of an integrated Society: A 21st century challenge*. Los Angeles: The Civil Rights Project/Proyecto Derechos Civiles at UCLA.

Perry, T. (2003). Up from the parched earth: Toward a theory of African-American achievement. In T. Perry, C. Steele & A. G. Hilliard, III (Eds.), *Young gifted and Black: Promoting high achievement among African-American students* (pp. 1–108). Boston: Beacon Press.

Planty, M., Hussar, W., Snyder, T., Provasnik, S., Kena, G., Dinkes, R., et al. (2008). *The condition of education 2008 (NCES 2008-031)*. Washington, DC: National Center for Education Statistics, Institute of Education Sciences, U.S. Department of Education.

Popper, M. (2000). The development of charismatic leaders. *Political Psychology, 21*(4), 729–744.

Richardson, J. W. (2005). *The cost of being poor: Poverty, lead poisoning, and policy implementation*. Westport: Praeger.

Richardson, J. W. (2006). The health and cognitive consequences of international child poverty. In C. C. Yeakey, J. W. Richardson & J. Brooks-Buck (Eds.), *Suffer the little children: National and international dimensions of child poverty and policy* (pp. 335–358). Oxford: Elsevier.

Richardson, J. W. (2007). Building bridges between school-based health clinics and schools. *Journal of School Health, 77*(7), 337–343.

Richardson, J. W. (2008). From risk to resilience: Promoting school – health partnerships for children. *The International Journal of Educational Reform, 17*(1), 19–36.

Richardson, J. W., & Harris III, J. J. (2004). Brown and its higher educational legacy: HBCUs and their "separate and unequal" standing. *The Journal of Negro Education, 73*(3), 365–378.

Riehl, C. J. (2000). The principal's role in creating inclusive schools for diverse students: A Review of normative, empirical, and critical literature on the practice of educational administration. *Review of Educational Research, 70*(1), 55–81.

Sanders, M. G. (2001). The role of "Community" in comprehensive school, family, and community partnership programs. *The Elementary School Journal, 102*(1), 19–34.

U.S. Department of Education – National Center for Education Statistics. (2004). *The condition of education 2004 (NCES 2004-077)*. Washington, DC: U.S. Government Printing Office.

U.S. Department of Education – National Center for Education Statistics. (2006). *The condition of education 2006* (Vol. NCES 2006-071). Washington, DC: U.S. Government Printing Office.

U.S. Department of Education – National Center for Education Statistics. (2008). *The condition of education 2008* (Vol. NCES 2008-031). Washington, DC: U.S. Government Printing Office.

Walker, V. S. (1996). *Their highest potential: An African American school community in the segregated south*. Chapel Hill, NC: The University of North Carolina Press.
Warren, M. R. (2005). Communities and schools: A new view of urban education reform. *Harvard Educational Review, 75*(2), 133–173.
Wilkinson III, J. H. (1979). *From Brown to Bakke: The Supreme Court and school integration: 1954–1978*. Oxford: Oxford University Press.
Wuenstel, M. (2002). Participants in the Arthurdale community schools' experiment in progressive education from the years 1934-1938. *Chula Vista, 122*(4), 759–770.

# 2

# The Efficacy of a Community Approach

> Educators need to know what happens in the world of the children with whom they work. They need to know the universe of their dreams, the language with which they skillfully defend themselves from the aggressiveness of their world, what they know independently of the school, and how they know it.
>
> – Paulo Freire (Freire, 1998, p. 72)

The term efficacy as it relates to advocating for a more holistic approach for children was purposefully selected because it denotes the power or ability to produce desired results. As demonstrated in Chapter 1, we are not producing admirable or equitable educational results for all of the nation's children. One has only to read a local newspaper article reporting the latest standardized tests scores to know that some children fare well in our public education system and others do not. Effectiveness hinges on our ability, as noted by Paulo Freire, to frame in-school interactions in ways that resonate with our students and incorporate relevant dimensions of their lives. Unfortunately, those who struggle in a standardized system that turns a blind eye to the factors directly and indirectly hindering learning will remain vulnerable in the absence of different educational approaches and those who will advocate on their behalf.

Advocacy is a delicate topic in educational circles because a notion prevails that teachers and principals ought to be apolitical or at the very least not actively engaged in the political arena. However, who is better informed than teachers and principals to advocate for a more effective model on behalf of children and their families? Given the time children spend in school, educators are well aware of risk manifestations. Unfortunately, the root causes of risk are most often outside a teacher's control. Educators live daily with the factors that trickle into classrooms across the country and steal potential from children, which is precisely why they and advocates for children should seriously consider the inherent value of the full-service community school approach. To embrace this shift one must not only understand the history of community schools, the landscape of contemporary education, but also how community schools might be an appropriate remedy.

Decades of research have concluded that children develop along multiple and interconnected domains. Disruptions to optimal physical, social, psychological, and emotional maturation potentially deter sensitive developmental sequences that impact learning (Blank & Berg, 2006; Richardson, 2008). By the time children enter school there are a host of school and nonschool factors that facilitate or diminish their learning readiness. We also know that when children are engaged, feel connected with their schools, and sense the relevance of what they are learning, their interest is sustained and naturally spills over into measurable academic outcomes.

Labels for both school-based and nonacademic deterrents to learning have come to be regarded as risk. A litany of risk factors associated with schools include but are not limited to curriculum rigor, teacher expectations, student – school connectedness, teacher preparation, and building safety. Nonschool domains are not only interconnected among themselves, but also have been verified to spill over and influence academic outcomes. Some of these detractors to learning are nutrition, stable and adequate housing, exposure to environmental toxins, general comprehensive health, parental engagement, and family and community resources. Ignoring the potential additive impact of disruptions to multiple developmental domains predictably reduces the likelihood that youth will optimize their innate potential (Beilke & Peoples, 1997; Benard, 2004; Blank & Berg, 2006; Freudenberg & Ruglis, 2007; Richardson, 2008).

Risk by definition tends to refer to any factor or a combination of factors that interfere with optimal development and is not relegated solely to children of low status. It is assumed to be universally experienced by children irrespective of ethnicity, gender, religion, socioeconomic status, and sexual preference. When wealthy parents fail to provide a loving, nurturing, and affirming environment their children are at developmental risk (Brendtro & Longhurst, 2005). Intellectually gifted youngsters are also at risk when stifled by boring curricula (Reiss, Colbert, & Hebert, 2005). However, when reflecting on those in greatest jeopardy, it is commonly assumed that poor children are exposed to sustained multiple risk factors that forecast academic and health difficulties with higher probability than their upper-income counterparts.

No dimension of a child's life is immune from risk resulting in researchers' calls for a holistic approach to its reduction. Furthermore, if children are subjected to one deterrent to development, it is highly likely they experiencing other risk factors concomitantly (Reiss et al., 2005; Richardson, 2008). Health markers of asthma, lead poisoning, and minimal access to care are emblematic of the types of housing families can afford. Housing location largely determines the type of schools children attend, and, as previously referenced, poor children are most apt to attend under-resourced public schools. Academic outcomes of youth in high-poverty neighborhoods (some of which are manifestations of health issues) also diminish the likelihood the cycle of poverty will be broken (The Annie E. Casey Foundation, 2007; Children's Defense Fund, 2006a, 2006b; Dorn, 2007; Douglas-Hall & Koball, 2006; Fass & Cauthen, 2005; Federal Interagency Forum on Child and Family Statistics, 2007; National Center for Children in Poverty, 2007a, 2007b; U.S. Department of Education – National Center for Education Statistics, 2006).

As you will note for the moment, race was left off the list of risk factors. I resist the presumption that race and ethnicity in and of themselves should be deemed a risk factor before people understand that such an idea often translates into deficit thinking about children at first glance. Race is a socially defined construct tied primarily to phenotype. Ethnicity on the other hand refers to group membership based on shared heritage and customs, which may be different from race. Certainly visual categorization of individuals may save time in record keeping; however, no child is immune from the bias (favorable or unfavorable) tied to their race.

More specific to Black youngsters, because of their race, expectations of their academic potential is less than that of other children and assumptions are that their performance on academic measures will be lackluster. The work of Claude Steele, Asa Hilliard III, Jerome Morris, and others confirms that students respond to the verbal and nonverbal expectations of their teachers (Hilliard III, 2003; Morris, 2008; Perry, 2003; Steele, 2003). In fact relative to resilience, the single greatest predictor of success is the expectation of the teacher (Benard, 2004; Peltzman & Jerald, 2006).

The power of expectations has been studied in depth by Claude Steele, who proposes that the nagging achievement gaps between groups of students are directly influenced by stereotype threat (Steele, 2003). Then (1940s) as now, Black students engage in an internal dialogue about the value of their effort in school compared to perceived benefits. As a result, why would students commit to achievement if they cannot predict success or have personally experienced a discounting of their effort? Why would they intellectually exert themselves if it is evident teacher expectations are that no matter how hard they try their results will be inferior to those of their White classmates? Why should they work hard when they have come to believe that no matter what they accomplish members of their group will be devalued (Perry, 2003)? This may seem a bit extreme; however, the experiences of persons of color (irrespective of educational level) being shadowed in retail establishments provides just one example of how young people become aware of society's negative perceptions of their race. Answers to these pertinent questions and observations create stereotype threat.

Stereotype threat has been defined as the phenomenon where negative attributions are ascribed to a particular group. Overt and subtle messages ascribing lackadaisical academic performance on certain students create an air of inevitability that children absorb. Self-fulfilling prophecies of this sort have been labeled stereotype threats. These negative expectations only matter because consciously or subconsciously target groups care about the phenomenon at hand (Steele, 2003). For example, Black youth are bombarded with messages about the likelihood they will not perform as well academically (particularly on standardized tests) as their White or Asian classmates. This premise is affirmed in literature, the media, and the expectations of teachers and possibly even parents. The fear that one might actually live up to that prediction creates a psychological burden that whether accurate or not can translate into a self-fulfilling prophecy, that is, a persistent achievement gap. A significant contribution of Steele's research is that in over a decade of empirical research he has confirmed that not only are all students susceptible to stereotype threat, but it is the most motivated and competent students who are most sensitive to the phenomenon. If negative expectations can assert themselves on the vulnerable psyche of children and youth, then so can positive affirmations.

When schools and communities create environments that are havens for "identity safety" threats can be mitigated. Stereotype threat in the reverse as tested by Steele and his colleagues demonstrated the power of affirming students' intellectual prowess as an elixir (Steele, 2003). Antidotes for risk, and by inference stereotype threat, produce the essential elements of resilience. Thus Steele proposes that if three domains are addressed effectively positive academic outcomes become more likely. They are shifts in pedagogical strategies and the relationships between teachers and students; adjustments to institutional practices and the context of the learning environment; and the individual student's responses to attendant changes in expectations from the broader societal predictions. He also argues that any changes from dominant norms and practices need to be explicit, demonstrative, and comprehensive. That is to say, students should be told (repeatedly) that they are being held to high academic standards, that they have the capacity to excel, and tangible support to excel is readily accessible.

The good news is that grounded in empirical findings, the above-noted approaches yield the results they claim, that is, turning the affirmative assertions about academic prowess into self-fulfilling prophecies (Steele, 2003).

So where does that leave us as we consider risk and resilience factors (now including race and ethnicity) that impact school environments and learning? First we should note that the contemporary discourse about Blacks and the "achievement gap" and its attendant causes is not much different than what was assumed to be true during the segregated era. Visions of dilapidated buildings, substandard textbooks, and teachers were and continue to be blamed for the disparate performance of Blacks as compared to their White age cohorts (Morris, 2008).

And now a word about the other side of the risk coin—resilience. Common across the generic definitions of resilience is the ability to defy negative predictions as a function of meaningful protective factor intervention(s) (Jensen & Fraser, 2006). First, a person has to be exposed to a degree of adversity that increases the probability for negative emotional, psychological, social, and/or behavioral outcomes (Bellin & Kovacs, 2006; Brooks, 2006; Kitano & Lewis, 2005; Reiss et al., 2005). Second, the quality of adaptation or the ability to mitigate the consequences of adversity varies; and third, protective or resiliency factors deflect negative outcomes (Bellin & Kovacs, 2006; Reiss et al., 2005). Stated another way, resilience is the ability to overcome adversarial factors that would typically predict failure, as well as to survive and recover from trauma (Brooks, 2006; Edwards, Mumford, Shillingford, & Serra-Roldan, 2007; Jenson, 2007; Kitano & Lewis, 2005). In its most inclusive form, "[r]esilience is an ecological phenomenon. It cannot be developed by sheer willpower within the at-risk person; it is developed through interactions within the environment, families, school, neighborhoods, and the larger community" (Brooks, 2006 p. 70).

Exposure to risk, taking resilience into account, does not necessarily destine children to failure. Some families and students have demonstrated their ability to "beat the odds" by creating support networks that create a safety net of services for children (Orthner, Jones-Sanpei, & Williamson, 2004). Thus, there is a need to acknowledge the inherent strengths families and communities have demonstrated, and also the social systems best positioned to promote similar outcomes (i.e., resilience) in children and youth (Orthner et al., 2004).

There have always been schools and families pre- and post-*Brown* who despite the prevalence of risk defied inferences predictive of failure. Harkening back to the days of JACS Black educators advanced counter-narratives as an antidote for stereotype threats. One prime reason for their success was the expectation that students had the intellectual capacity to excel academically. Another was the purposeful assembling of academic support extending beyond school walls into churches, barbershops, stores, and other venues. Clues as to how this was possible in the days of James Adams Community School (JACS) come from the resilience and protective factor literature.

Childhood resilience manifests itself as personal flexibility, adaptability, motivation, social responsibility, and creativity. Institutions in addition to individuals can nurture resilience through responsive atmospheres and well-run organizations (Mandleco & Perry, 2000). Educational resilience has specifically been defined as "the heightened likelihood of success in school and other life accomplishments despite environmental adversities brought about by early traits, conditions, and experiences" (Reiss et al., 2005, p. 111).

Resilience is a dynamic rather than static phenomenon and is nuanced by virtue of culture and community, thus suggesting the integrated structural and programmatic processes in community schools remain fluid (Bellin & Kovacs, 2006; Benard, 2004; Mykota & Muhajarine, 2005; Reiss et al., 2005; Richardson, 2008). Interventions designed to support youth resilience need to take into account that singular interventions and one-size-fits-all programs will not yield the greatest results in children. As youngsters grow and familial needs change, so do the structures needed to nurture resilience. That said, the educational establishment cannot function as a protective agency in isolation.

In an article concerned with school – community partnerships, Lee-Bayha and Harrison (2002) posit, "The best of teaching cannot always compete successfully with the challenges many students face outside of school" (p. 1). What becomes clear in a survey of the literature is that possessing the capacity to bounce back from challenges typically predictive of failure requires a combination of personal attributes, positive relationships, and institutional supports. Thus most often, internal and external strengths in tandem aid in overcoming challenges that manifest themselves in dynamic, contextual, culturally influenced ways. Furthermore, resilience is highly dependent upon relationships and programs at the places children spend the bulk of their time, school and home.

There is ample evidence of the links between the efficacy of community – school collaborations fueling calls for multidisciplinary interventions in general, and in support for resilience in particular. As a result, resilience has been identified as an ecological phenomenon because an ability to surmount challenges is not believed to exist solely within a persona. In addition, because risk and resilience factors do not occur in isolation and they interact perpetually, there is a call to think about

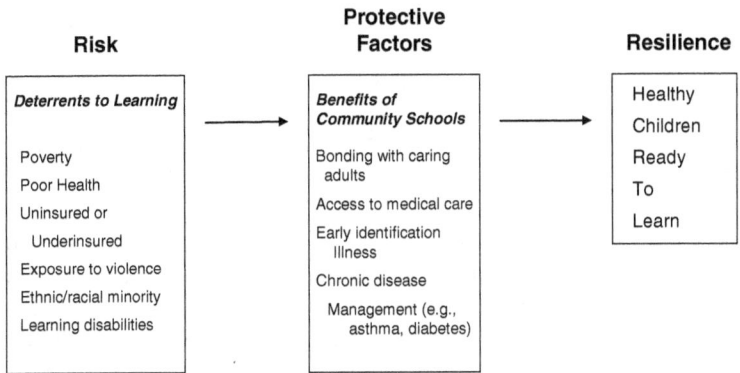

**Figure 2.1** From Risk to Resilience.
*Source:* Geierstanger, Amaral, Mansour, & Walters, 2004; Richardson, 2007; Rounds & Ormsby, 2006.

childhood interventions at critical junctures using an approach inclusive of families, social systems, programs, and communities (Bellin & Kovacs, 2006; Brooks, 2006; Mandleco & Perry, 2000).

Juxtaposed between risk and resilience researchers have inserted protective factors as those resources that minimize or mitigate risk (Figure 2.1). Individual physiological, emotional, and intellectual characteristics such as general health, IQ, and coping ability are considered internal contributors to resilience. Families and other organizations, such as schools and healthcare agencies, are considered external protective factors (Mandleco & Perry, 2000; Richardson, 2008).

Relative to converting risk to resilience, community schools can be immensely helpful particularly if children are economically challenged or live in regions underserved by the traditional social welfare agencies. Partnering of schools, agencies and community-based businesses and organizations can have a profound effect on learning outcomes. Full-service schools can be protective factors against risk for not only students, but their families as well. When schools are the site of a web of services, programming and activities valued by the community, they function as agents against risk for all who are involved (Benard, 2004). Protective strategies should include comprehensive services and adequate dosages of interventions with sufficient intensity, be socioculturally relevant, and implemented by well-trained staff (Edwards et al., 2007; United Nations Educational Scientific and Cultural Organization [UNESCO], 2002).

There are many vantage points from which an investigation of the integration between and among risk, resilience, and protective factors impacts learning. Framing childhood initiatives as holistic endeavors is well supported in theoretical frameworks advanced by scholars. The work of Blank, Dryfoos, Dewey, and others also resonates with calls for schooling to be informed by, be integral to the fabric of communities, and confirm that schools working in concert families and communities are critical to student success (Blank, 2004; Dewey, 2001/1915; Dryfoos & Maguire, 2002; Jensen & Fraser, 2006; Kronick, 2005; Sanders, 2001). Given the plethora of theories that could inform this study, there was a need to narrow the pool to those that embody the best thinking about community – school engagement. As a result, the Ecology of Human Development (EHD) first advanced by Urie Bronfenbrenner is central to this work.

The EHD framework posited by Bronfenbrenner (1979) has been widely used to explain why some children might be more resilient than others. It has not, however, been pervasively applied to validate the value of community schools as a way to improve the lives of children. However, as risk and resilience factors do not occur in isolation and they interact perpetually, it is appropriate to use an ecological lens to think about childhood interventions at critical junctures using an approach inclusive of families, social systems, programs, and communities (Bellin & Kovacs, 2006; Brooks, 2006; Mandleco & Perry, 2000).

### The Ecology of Human Development and Community Schools

The EHD framework posited by Bronfenbrenner (1979) is an ecological theory presuming certain things: it requires more than a singular vulnerability factor to be at work in a student's life, and it assumes that interventions at multiple levels (personal, family, community, institutional, and broader society) take place concurrently (Edwards et al., 2007).

Three fundamental concepts undergird the framework. First, the Ecology of Human Development asserts persons as dynamic entities upon whom environments exert influence. Second, individuals and their environment reciprocally interact, creating a need to accommodate one another in the developmental process. Third, environments and developmental processes are not one-dimensional, but rather extend between settings emanating from narrow to broad contexts (Bronfenbrenner, 1979). Each of these assumptions aligns with the characteristics of protective factors and resilience already discussed. Concentric circles representing micro-, meso-, exo-, and macrosystems are most often used to signify the perpetual interaction of multiple contexts (see Figure 2.2).

Central to development of child – adult relationships are exchanges that occur between youth and significant others such as parents and teachers. Microsystems (i.e., one-on-one exchanges) take place in settings such as homes, day-care facilities, and schools, and are where childhood

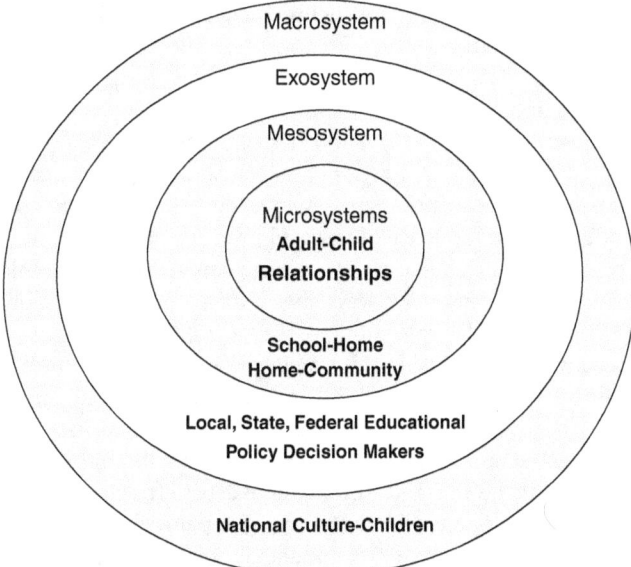

**Figure 2.2** Ecology of Human Development.
Source: Bronfenbrenner, 1976, 1979.

experiences influence development (Bronfenbrenner, 1976, 1979; Mertensmeyer & Fine, 2000).

Interrelations between two or more settings such as home and school, or home and community organizations are criteria for mesosystemic influence. In schools, a developing child is forced into a new setting requiring an adjustment to new norms, behaviors, and expectations. Mesosystems like microsystems are settings where children spend most of their time. By virtue of the hours spent in school for example, opportunities to foster relationships and implement interventions that shape early learning, health, and behavior exist (Mertensmeyer & Fine, 2000). Another way to conceptualize a mesosystem would be a system of microsystems (Bronfenbrenner, 1979).

Settings that extend their influence to a child but do not directly involve the child are considered exosystems (Bronfenbrenner, 1979). Formal agencies and political bodies charged with distributing resources and dictating practices in schools at the federal, state, or local level such as departments of education and school boards are examples (Brendtro, 2006; Bronfenbrenner, 1976; Renn, 2003).

Finally, a macrosystem speaks to the larger culture, which heavily influences all other systems. For example, the global economy, a national culture, and the belief systems and ideologies that support them indirectly influence exo-, meso-, and microsystems (Brendtro, 2006; Bronfenbrenner, 1976, 1979; Renn, 2003).

Each of the systems (micro-, exo-, meso-, and macro-), depicted as concentric circles, imposes pressure on youth simultaneously as they mature. Individual development, change, and transition occur normally and naturally as a function of interaction between and among these systems and, as such, the EHD model proves its utility as a heuristic approach when pondering the influence of risk and protective factors on resilience. Risks, or—in Dr. Bronfenbrenner's vernacular—"disruptive ecologies," manifest themselves in many ways. Reaction to these disruptions creates "dis-ease" with the environment (Brendtro, 2006; Bronfenbrenner, 1976). Terminology selected in this framework rejects the notion that children possess deficits; rather children respond to deficits in the various dimensions of their environment (developmental disabilities notwithstanding). Transactions occur between children, adults, and systems that create behavioral, health, and developmental patterns that either facilitate "normal" development or thwart normal patterns. Given our focus on resilience, disruptions to developmentally sensitive interactions create risk that creates a need for protective factors to counteract them.

Application of a resiliency lens to the EHD framework would posit "the most powerful interventions with children and youth are those that seek to build a supportive ecology around a child" (Brendtro, 2006, p. 165). As an applied example of Bronfenbrenner's work, the problem with risk and resilience is not so much a child-based problem as a need to introduce protective factors into high-risk ecologies. Applied to education he noted,

> Whether and how people learn in educational settings is a function of sets of forces, or systems, at two levels: The first comprises the relations between the characteristics of the learners and the surroundings... The second encompasses the relations and interconnections that exist between these environments. (Bronfenbrenner, 1976, p. 5)

That said schools and the services offered within them are prime locales for protective interventions.

EHD explains why schools are important and should not be thought of as compartmentalized from the rest of a child's life. Home, school, relationships with adults, and local, state and federal political structures all influence children simultaneously. As politicians and agency officials prioritize the distribution of educational resources (exosystems) they reveal who and what is highly valued, and conversely, who and what is not. Therefore some disruptions to optimal development are increasingly acknowledged as symptomatic of intersections between political, economic and social networks that envelope the world in which children live (Buttery & Anderson, 1999).

EHD extended further to incorporate findings of additional scholars would suggest that the ecologies of children differ based on the environments they find themselves in, particularly in light of race, ethnicity and class (Ames & Farrell, 2005; Beilke & Peoples, 1997; Hilliard, III, 2003; Newman & Chen, 2007). Availability of quality adult – child relationships (microsystems) and school and community assets (mesosytems) vary greatly as data supporting the prevalence of school resegregation and the attendant quality of these educational environments would suggest (Orfield, 2009; Planty et al., 2008). Simultaneously, the scope of the national instructional delivery systems and accountability pressures has by and large reduced pedagogical foci in many schools to a standardized approach to curricular implementation. The problem with that strategy is that it ignores the real risk exposure of students and misses opportunities to serve as protective factors. It is subsequently no surprise that achievement gaps persist.

The notion that pedagogical strategies working well for Caucasian or middle-income children are good for all children needs to be dispelled. If we continue to ignore the idiosyncratic needs of children and their communities in favor of an inflexible public education system insensitive to community idiosyncrasies, we will continue to see the results we have reaped—academic disparities. Perry (2003) argues that since learning is contextual, effective learning environments would be those that honor the social, emotional, cognitive, cultural, and political sensibilities of the communities served.

Many articles focus on the individual characteristics that lend themselves to resilience. Far fewer articles consider the role of systems as protective factors for children. This text began making the argument that children of poverty are in greater need of systemic interventions that support optimal learning readiness and health because their families do not command as many resources as their upper income counterparts. A community school ethos inherently takes into account the influences of micro-, meso-, and exosystems in the creation of a protective environment. With poor and near-poor children most apt to be uninsured or underinsured, and also quite likely attending a school lacking requisite resources, intuitively partnerships hold the possibility of rebuilding shields around childhood potential (Newman & Chen, 2007).

In the African American community two institutions have historically "stood in the gap" as protective factors: churches and schools. While some researchers argue nurturing achievement in the African American community is distinctive (and by inference others as well), community schools by definition adapt to neighborhood contexts. As such they would be the place where incongruence between the myths a larger society might hold of you (macrosystem narrative) and what you know to be true about your value and potential can be nurtured (microsystem counter-narrative). Following this line of thought further, if schools are to lend protection to youth, one model that has proven over time to be effective is that of community and full-service schools. That said, all community schools are not the same and so there is a need to consider an implemented model that lends itself to informing a broad potential audience.

How is it that segregated community schools functioned as protective agents supporting what contemporary literature defines as resilience? In short, it was viewed as a matter of necessity, in part because of unequal educational financing and segregation practices, that community schools pre-*Brown* functioned along with other agencies to form a dependable safety net for students. Black schools and their communities in the South commonly banded together to ensure students had the supplies and assistance they needed to be successful. When money was in short supply parents and community members would volunteer their time, skills, vehicles, and even physical labor to bolster the teaching – learning exchange. Consider the example of Paul Lawrence Dunbar High School in Little Rock, Arkansas, from 1930 to 1955. At the time it was the only high school available to Black students. Teachers and administrators were part of the community so they not only worked there, but they went to church, participated in civic life, and as such knew their students, families and resources they could tap for schools. In Tuscumbia, Alabama, another example has recorded as a symbiotic relationship between the local Black high school and community (Morris, 2008).

Resilience literature has confirmed what teachers in segregated Black schools intuitively knew: adult expectations heavily influence youth. High expectations as per the exhaustive synthesis of resiliency literature sponsored by WestEd merits the designation of a protective factor. Common across schools demonstrating their ability to facilitate the closing academic performance gaps (pre- and post-*Brown*) is the refusal to "dumb down" curricula. Instead these schools by virtue of teacher – student interactions, pedagogical practices, and other factors maintain that if students believe that the adults around them are convinced of their abilities, they tend to manifest their intellectual prowess in ways educational institutions tend to measure success (Benard, 2004). Risk in the African American tradition was not viewed as a predictor of school failure; instead needs were met as best as could be managed by programming and other resources.

The power of the verbalization of expectations brings us back to the scholarly investigations of Dr. Steele (2003), who recounts that expectations have much to do with measurable student outcomes. Consider the single greatest protective power of schools is housed in the expectation that all students have the capacity to excel academically. Now multiply the power of that teacher's expectation with a triangulated version that shows up in every sector of a child's life and consider the potential impact on a youthful psyche. One's pastor, barber, doctor, local candy store owner, and parents not only verbally value education and remind you of your potential, but demonstrate how important education is by contributing their time to your clubs, classes, festivals, basketball games. That is what life was like at JACS and how T. J. and Anita Anderson built, in collaboration with many others, a nurturing island of protection that propelled students beyond the narrow life expectations their economic or racial station might otherwise predict.

EHD and resiliency theories, while thought provoking, tend not to translate easily into tactics community school advocates can employ. Neither do they aid in the constructing of measurable outcomes of a more comprehensive and holistic school experience for students. If we are to look back at JACS and extract relevant lessons forward, we need a road map of sorts to guide the historical analysis using the contemporary lexicon. Guidance also needs to be flexible enough to honor the idiosyncratic nature of full-service schools that seek to be culturally and community relevant. The Highly Effective Community School (HECS) model was developed using findings from leadership, resilience and community school literature, and JACS data in an effort to support not only this evaluation but others to come (Figure 2.3).

Divisions of the model focus on four categories: leadership, partnerships, community school organizational development, and benefits of a highly functioning community school. The figure also conveys the interdependence of the first three categories between and among each other and to beneficial outcomes. The relationships between leadership, collaborative partners, and organizational development build on each other over time producing, in a best-case scenario, a sustainable successful institution. That is not to say that attempts to improve organizational capacity alone, for example, could not positively influence children and their families. The point here is to identify the optimal integration that as per the literature yields the best student and community outcomes. The subsequent discussion provides a rationale for the model and descriptions of how it relates to this assessment of JACS.

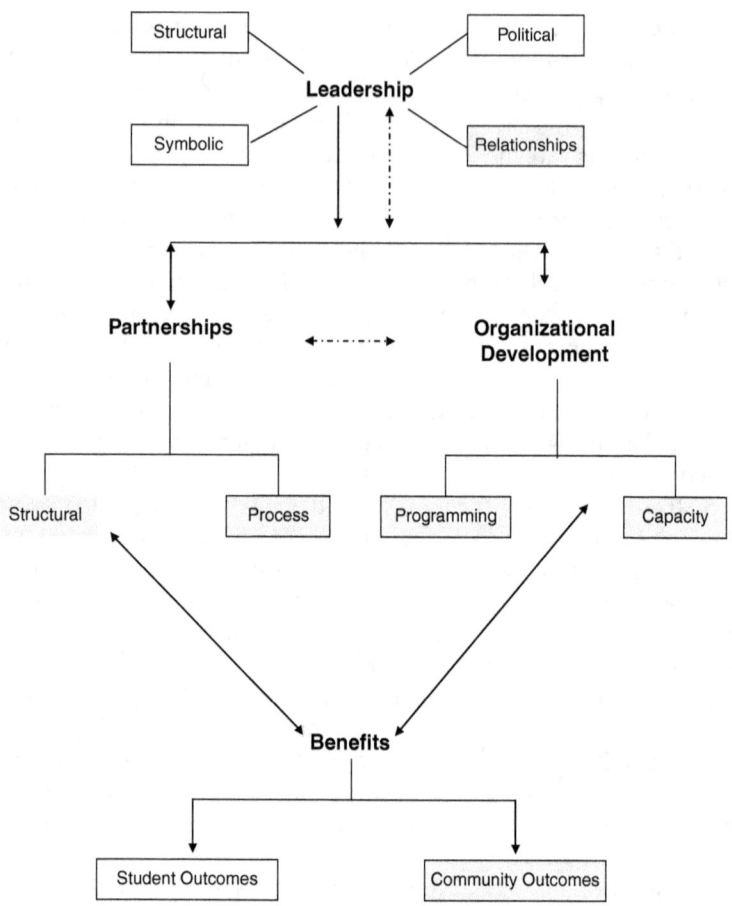

**Figure 2.3** Highly Effective Community School (HECS) Model.

*Leadership*

Proper attention to semantics would caution us to perpetually make distinctions between leaders and leadership. A leader is an individual who is responsible for the outcomes of the entity over which they have jurisdiction. S/he has innate qualities that inform her/his interactions with subordinates, and her/his training, education, and experience inform decision making. The leader behaviors exhibited with staff and general public are what I refer to as leadership. Leader qualities and the markers of effective leadership are the focus of researchers across numerous disciplines.

There appears to be no consensus on whether leaders are born or can be trained; or whether good managerial instincts are innate or behaviors that can be taught. Though the mystery has yet to be completely unraveled the influence of dynamic leaders cannot be denied. Leadership in this text is not considered synonymous with supervision, which speaks to influencing individuals or groups singularly through punishments or contractual obligations (Jago, 1982). Instead, leadership is much more fluid. A comprehensive definition that includes both the idiosyncratic leader persona and leadership is noted below:

> Leadership is both a process and a property. The process of leadership is the use of noncoercive influence to direct and coordinate the activities of the members of an organized group toward the accomplishment of group objectives. As a property, leadership is the set of qualities or characteristics attributed to those who are perceived to successfully employ such influence. (Jago, 1982, p. 315)

The individual characteristics and preparation are ascribed to an individual (i.e., "property"), while the processes associated with leading are interpreted here as leadership.

Jago (1982) attempted to further illuminate his perspectives on leadership by dividing empirical studies into categories. Demarcations include those focused on innate characteristics of leaders; leadership skills that could be taught; and conditions under which either innate or learned skill yield effective results. The results of his study and others lead us to the conclusion that try as we might to define leader and leadership qualities that guarantee success, leadership styles and leaders that demonstrate both competence and charisma have an indescribable "it" factor (Jago, 1982; Kouzes & Posner, 1995; Popper, 2000). That said there are some markers for successful leadership that can be identified in general and for school principals in particular.

Relative to schools, their hierarchical governance structure lends itself to a need for tremendous coercive influence by educational leaders. Multiple federal, state, and local political bodies dictate school expectations and pass down rules and regulations principals are obliged to honor. Principals have long been considered pivotal to school success. Interestingly, successful schools have more often than not been linked to the leadership styles and qualities of principals than to specific policies and the power exercise by governing bodies. Principals have been blessed and cursed with more responsibility than they could possibly manage (Camburn, Rowan, & Taylor, 2003). They are to be instructional, disciplinary, and managerial leaders. Yet more and more studies admit that optimally leadership in schools is tied to the fostering and empowerment of networks and the tenacity of principals to know when to be autocratic and when to be democratic; when to be hands-on versus hands-off; and when to be task- as opposed to relationship-oriented (Camburn et al., 2003; Jago, 1982; Rice & Harris, 2003).

Four principal supports of effective leadership are depicted in Table 2.1 (and Figure 2.3) as structural, human relations, political, and symbolic. These categories were advanced by Rice and Harris (2003) and adapted our investigation of the leader characteristics and leadership of T. J. Anderson, principal at JACS.

Structural dimensions of leadership pertain to the formal position appointment, academic and professional preparation, and skill planning and implementing a vision. Human relations refer to the relationships the leader has, nurtures, or creates with people and organizations. Political attributes of effective leaders are tied to power relationships, formal and informal influence, and an ability to commandeer resource to support the proposed structure. Finally, symbolic leadership qualities include charisma, personal commitment, tenacity, and creativity, and ability to articulate the values and vision of the organization (Rice & Harris, 2003). Literature supports definitions of different types of charisma as personalized and socialized. Socialized charisma, however, is most relevant

**Table 2.1** Leaders and Leadership

| Structural | Political |
|---|---|
| – Understand community challenges | – Influential with community power brokers |
| – Exemplary academic preparation | – Influential with traditional and nontraditional power brokers |
| – Ability to employ strategies appropriate to the community context | – Defies professional turf |
| – Encourages linking life experiences to curricula | – Shares power and information |

| Human Relations | Symbolic |
|---|---|
| – Perceived as open/welcoming | – Possesses charisma |
| – Adept at working with interpersonal dynamics | – Demonstrates personal commitment |
| – Culturally competent | – Clearly articulates a vision |
| – Targets strategic partnerships | – Demonstrates an ability to motivate others |

*Source*: Jago, 1982; Popper, 2000; Rice & Harris, 2003.

here as it refers to those who lead through genuine regard for feeling and rights and then using his/her station to empower "followers" (Popper, 2000). Structural and symbolic designations are tied to individual leader qualities. Political and human relations represent leadership dimensions. In a practical sense the leader of a community school drives the initiative. S/he has prepared adequately for the position and can articulate in a compelling way the vision for the community school (symbolic and structural). Given the work involved and the numbers of persons and entities needed to bring a community school vision to reality, the principal needs to be focused, deliberate, and persistent. They need to understand the mores, values, and community assets of the families served by the school and be astute in identifying potential partners (structural and human relations).

As leaders, principals need to readily adjust leadership styles to suit each situation and command the influence to facilitate consensus among those whose decisions aid or deter progress (political and human relations). Trust is the thread that holds all of the plans and work together, because if the leader is perceived to be authentic in their commitment they are more likely to attract similarly committed individuals and groups (symbolic). Another intangible quality of a community school leader would be a willingness to take calculated risks (symbolic) (Blank, 2004; Dryfoos, Quinn, & Barkin, 2005). In the words of Dryfoos, "The ship in the harbor is safe, but that is not what ships were made for: a ship was made to go out into the sea and face risks" (Dryfoos et al., 2005, p. 131). Embarking on the creation of a full-service school is a risky departure from traditional practice.

Once a leader is in place as a principal, then partnerships should be explored to support the community school programming that extends beyond the traditional school day. Furthermore, partnerships are particularly important in neighborhoods that by any measure might be considered disadvantaged. As leadership has many definitions, so does partnership. As an attempt to ensure clarity, the working definition of a school – community partnership is as follows: "connections between schools and community individuals, organizations, and businesses that are forged to promote students' social, emotional, physical, [economic] and intellectual development" (Sanders, 2001, p. 20).

*Partnerships*

As schools moved philosophically toward stand-alone institutions they risked divorcing themselves from valuable community resources to varying degrees. However, taking into account Bronfenbrenner's theory, it is counterproductive to isolate spheres of influence on children when their impact is exacted concurrently. Furthermore, optimizing the intellectual potential of children does not occur only between the designated hours of instruction, but rather *en route* to school and home and all activities in between. Resources and influences beyond the school walls are inextricably linked to what happens inside school. Thus, partnerships between school personnel and community members are a valuable way to reinforce and enhance the learning experience (Buttery & Anderson, 1999; Harris & Hoover, 2003; Sanders, 2001).

"Community" in community and full-service schools could have any number of meanings. The community consists of parents, guardians, and families of students, those in the neighborhood who do not have children in attendance at the school, businesses, houses of faith, and political actors who influence what can and cannot happen on school grounds. School – community partnerships are connections that have been forged for the purpose of promoting the social, emotional, physical, psychological, and intellectual development of students (Sanders, 2001).

Inherent in this broad definition is a wide range of potential resources as well as conceptual priorities for the school programming. Harnessing the potential benefits of collaborations whether initiated by educational professionals or parents, for example, should begin with mutually agreed upon agendas, role designations (who will do what), power sharing, as well as formal and informal communications pathways to ensure the integrity of the task(s) at hand.

Partnerships should not be pursued just for partnerships' sake, but rather conceived through a mutual value for educational excellence, the benefits from building on the respective expertise, and commitment of both parties (Britton & Britton, 1970). Decisions to embrace partners must be coupled with a willingness to work hard at solidifying meaningful and sustainable engagement (Dryfoos, 2003). Leaders and those they charge with building collaborations need to be cognizant of partner language, mores, and priorities. In this way joining with schools is conceived not as additional work, but rather is in keeping with the existing priorities of said partners.

Chapter 1 indicated that each community school is unique because the assets and needs of each neighborhood surrounding the school vary. A moment is required to consider the impact of diversity, cultural competence, and cultural humility in identifying potential partners. Diversity in its simplest form illuminates difference between the school-building faculty, principal, staff, and the students they support. Cultural competence tends to refer to a set of skills designed to explore the existence of differences in an effort to appreciate the influence of bias, assumptions, and expectations of engagement. Cultural humility goes a step beyond cultural competence because it is a process by which individuals are perpetually self-reflective and critical of their understanding of others. The goal in practice is to identify similarities and differences between oneself and others in an effort to create a respectful partnership (Tervalon & Murray-Garcia, 1998). In order to meaningfully engage individuals and groups to participate fully in the educational process differences should be honored and incorporated into the functioning of the school where appropriate.

Partnerships are separated into two categories: structural and process. Structural aspects of partnerships answer the question, who are the partners? Strategic partners pervasive in the community school literature are school administrators, staff, faculty, community-based agencies (such as churches), parents, funders, local businesses, students, and other social service organizations (such as medical facilities) (Blank, Melaville, & Shah, 2003; The Children's Aid Society, 2001; Dryfoos, 2002). Collaborative processes, the second partnership category, refers to what partners do and how partners influence and are influenced by the school.

The most important partners educators can have are parents and studies have supported the positive influence of parental involvement on student success (Hardaway & McLoyd, 2009). However, in traditional schools, both by virtue of race or income, African American parents can be made to feel unwelcome, and assumptions made about their ability and willingness to support the life of the school. Theorists have argued that this friction between families and schools, particularly in the neediest school districts, is a function of educators not valuing the skills, resources, and culture of parents (Hardaway & McLoyd, 2009). Community schools by their definition seek to reverse this trend and invite parents in as equal partners in the creation of learning environments. Some good initial partnership activities include mentoring, tutoring, donations of supplies, and apprenticeships, all of which have led to improved academic outcomes and enhancing participants' connectedness to the school (Sanders, 2001).

It is easy to overlook the cast of contributors to learning in schools who by virtue of their employment may play direct or ancillary roles in support. Much attention was paid to the role of the principal earlier, and here we consider the critical contributions of other members of the school-building community. One of the most consistent findings in the resilience literature is the power of teachers as protective agents against risk (Benard, 2004). Teachers, maintenance and cafeteria staff, secretaries, security guards, counselors, and school nurses all play important roles in the development and sustainability of community schools. In the absence of their buy-in as partners the vision cannot be realized. Thus, while their professional affiliations require certain levels of engagement, the commitment required to elevate a school to the full-service stratum requires teachers and staff agree to invest more time and energy than their job descriptions might explicitly articulate.

Potential collaborators with community schools can also be social service agencies that establish offices in or near the school such as school-based health-care clinics. Businesses can lend financial support as patrons of programming and also become sites where students are interns or apprentices. Houses of faith can lend aid in many ways without the imposition of their specific belief system on schools. For example, they can adopt classrooms and collect supplies for students and teachers. Grassroots advocacy organizations can speak on the schools' behalf at political events. The list is limited only by the imagination of the leader and those s/he empower to devise a web of full-service school support. When carefully considered, there will be long list of potential partners. The more partners are meaningfully engaged and empowered to co-create relationships with school faculty and staff, the more engaged they become. Yet, even the best-intentioned collaborative plans will fail if they are not concomitantly supported by organization development.

*Organization development*

Organizations are in a practical sense groups of individuals who work together to accomplish a common set of goals. It has also been proposed that organizations of all sorts enhance their capacities by virtue of partnerships. Functioning independently is an inefficient way to meet organizational goals (Buttery & Anderson, 1999). Though not always considered an organization in a business sense, community schools are at their best a collaborative neighborhood problem-solving enterprise, organized around the priorities and needs of the community (Britton & Britton, 1970).

The HECS model designates two organizational distinctions: programming and capacity. Programming considers the organizational goals and what after- or before-school activities will be made available and for whom. Programming as a result of needs and community asset assessment should be informed by partners and the general community so that offerings are relevant and culturally appropriate. The types of programming indicative of a community school can comprise but is not limited to extended learning opportunities, parental involvement and support, community asset building, innovative and rigorous curricula, and student health promotion (The Children's Aid Society, 2001). This collaborative effort presumes a degree of power sharing with partners. Adherence to the full-service community school model in its purest form requires community-directed programming as opposed to schools establishing programming for a neighborhood (Blank et al., 2003; Clapp, 1971; Dewey, 2001/1915; Dryfoos, 2002). This requires a level of cultural humility and joint crafting of an organization built on community priorities and an honest assessment of the resources needed to sustain the organization.

The school administration, faculty, and staff in the process of considering programming must identify their strengths and acknowledge weaknesses in order to employ strategic partnerships to enhance capacity. Capacity in this context also includes how the programming is implemented to include staffing, funding, facilities, and so on. Furthermore, organizational capacity considers how strategic partnerships will advance the vision of the school, how community needs will be assessed, and how mutually agreed upon innovations will be funded and staffed. Fundamentally, the organization (full-service school) is viewed as constructed to meet the needs of children rather than children needing to adjust to fit into a predetermined organization. A word about resources and assets is in order at this juncture. Any support of a community school can be categorized as a resource. It is not limited to money alone, but rather can include volunteers' time, classroom supplies, and donated facilities and expertise.

Determining programming and organizational capacity of a community school is the result of a dynamic and emergent process. Facility limitations may make certain requested offerings impractical and compromises need to be struck. Partners may bring expertise and resources to the school enhancing the physical plant. The leader may then parlay innovations into press releases, as one example, that expand the audience size familiar with the contributions originating at the school site. The principal advantages of the full-service school model, fluid organizational development, leadership, and engaged partners, are the benefits that accrue not only to students but also the surrounding region.

*Benefits*

When leadership, partnerships, and organizational structures are in place the desired outcomes for students and communities are both tangible and intangible. Tangible student outcomes are grade completion and promotion, performance on standardized tests, degree attainment, building community assets, and better use of the facility (Blank et al., 2003; Peebles-Wilkins, 2004). The intangibles are more difficult to measure, but are just as significant, such as a commitment to lifelong learning and community service, and community pride (Blank et al., 2003; The Children's Aid

Society, 2007; Dryfoos, 1994; Dryfoos & Maguire, 2002; Dryfoos et al., 2005; Peebles-Wilkins, 2004).

Chapter 1 devoted substantive attention to the benefits that community schools offer students. Something easily overlooked are the benefits that can accrue to partners. Parents for the most part want their children to have access to opportunities. Though they desire their children to be successful in school, they may lack an understanding of how to help their youngsters. They might be inclined to volunteer in the school, but if the school is not welcoming they may withdraw. The literature is clear that parental involvement in school enhances student motivation and breaks down barriers between the school and community. Students whose parents are engaged in the life of the school tend to have higher achievement, demonstrate motivation, and are less apt to experience discipline problems (Balfanz, Fox, Bridgeland, & McNaught, 2009; Keith, 1999). Programming can also offer parents desperately needed social and professional support as well as creative outlets.

Teachers as partners benefit from a community school setting. They can devote more of their energies to innovative pedagogical strategies because of the other needs of students being addressed elsewhere. Companies can benefit because aside from the positive public relations associated with aiding schools, they also ensure the next generation of employees have the requisite workforce skills in demand.

Though Figure 2.3 displays leadership, partnerships, and organizational development in a linear fashion, be assured that the interactions are fluid, compounding, and interactive, as represented by the bidirectional arrows. The unequivocal starting point is the leader's structural, human resource, political, and symbolic aptitudes. Once the leader is well established, the school and community partnerships are attracted. Partners meaningfully engaged in school life will attract others as benefits begin to accrue that may be tangible such as school-based products or intangible, that is, positive public relations and visibility. As resources directly or indirectly flow from partners to the school, the organizational capacity structurally increases in both the programming that can be offered and how it is paid for. Students are not only told they have potential, but they experience a manifestation of caring adult sentiments reinforced by the actions and curricula of teachers and dedication of the community to their success, and as such ties to community identity and a sense of responsibility to excel is nurtured. Stated another way, they learn to combat stereotype threats because they are armed with a counter-narrative demonstrated by the adults in their world.

Now that the model guiding the analysis has been articulated, theoretical underpinnings and benefits of community schools expounded, and the similarities between pre- and post-*Brown* student bodies established, we now turn our attention to JACS. Data culled from interviews, school and personal documents, and newspaper articles populate the HECS categorical evaluation and each will compare JACS data to contemporary accounts of best practices. Tangible outcomes 50 years after the close of the school are difficult to come by. However, the intangible evidence of JACS's enduring influence is evident in the interviews and correspondence received. Subsequent chapters expound on the particular divisions of the model (leadership, partnerships, organizational development, and benefits) followed by a reconstruction of the model in its entirety.

### References

Ames, B. D., & Farrell, P. (2005). An ecological approach: A community-school strategy for health promotion. *Journal of Family and Consumer Sciences, 97*(2), 29–35.
The Annie E. Casey Foundation. (2007). *2007 kids count data book*. Baltimore: Annie E. Casey Foundation.
The Children's Aid Society. (2001). *Building a community school* (3rd ed.). New York: The Children's Aid Society.
The Children's Aid Society. (2007). *Community schools*. Retrieved October 25, 2007, from http://www.childrensaidsociety.org/communityschools.
Balfanz, R., Fox, J. H., Bridgeland, J. M., & McNaught, M. (2009). *Grad Nation*. Retrieved February 21, 2009, from http://www.americaspromise.org/uploadedFiles/AmericasPromiseAlliance/Grad_Nation/GradNation_020509.pdf.
Beilke, J. R., & Peoples, G. (1997). Failure to thrive syndrome (FTTS): Predicting educational failure at the middle school level. *Education, 117*(4), 512.
Bellin, M. H., & Kovacs, P. J. (2006). Fostering resilience in siblings of youths with a chronic health condition: A review of the literature. *Health and Social Work, 31*(3), 209–216.
Benard, B. (2004). *Resiliency: What we have learned*. San Francisco: WestEd.
Blank, M. J. (2004). How community schools make a difference. *Schools as Learning Communities, 61*(8), 62–65.
Blank, M. J., & Berg, A. (2006). *All together now: Sharing responsibility for the whole child*. Washington, DC: Association for Supervision and Curriculum Development.
Blank, M. J., Melaville, A., & Shah, B. P. (2003). *Making the difference: Research and practice in community schools*. Washington, DC: Coalition for Community Schools, Institute for Educational Leadership.
Brendtro, L. K. (2006). The vision of Urie Bronfenbrenner: Adults who are crazy about kids. *Reclaiming Children and Youth, 15*(3), 162–166.
Brendtro, L. K., & Longhurst, J. E. (2005). The resilient brain. *Reclaiming Children and Youth, 14*(1), 52–60.

Britton, J. O., & Britton, J. H. (1970). Schools serving the total family and community. *The Family Coordinator, 19*(4), 308–316.
Bronfenbrenner, U. (1976). The experimental ecology of education. *Educational Researcher, 5*(9), 5–15.
Bronfenbrenner, U. (1979). *The ecology of human development.* Cambridge, MA: Harvard University Press.
Brooks, J. E. (2006). Strengthening resilience in children and youths: Maximizing opportunities through the schools. *Children and Schools, 28*(2), 69–76.
Buttery, T. J., & Anderson, P. J. (1999). Community, school, and parent dynamics: A synthesis of literature and activities. *Teacher Education Quarterly, 26*(4), 111.
Camburn, E., Rowan, B., & Taylor, J. E. (2003). Distributed leadership in schools: The case of elementary schools adopting comprehensive school reform models. *Educational Evaluation and Policy Analysis, 25*(4), 347–373.
Children's Defense Fund. (2006a). *Black child health fact sheet.* Retrieved September 24, 2007, from http://www.childrensdefense.org/site/PageServer?pagename=policy_ch_blackfactsheet.
Children's Defense Fund. (2006b). *Latino child health fact sheet.* Retrieved September 24, 2007, from http://www.childrensdefense.org/site/PageServer?pagename=policy_ch_latinofactsheet.
Clapp, E. R. (1971). *Community schools in action.* New York: Arno Press & The New York Times.
Dewey, J. (2001/1915). *The school and society & the child and the curriculum.* Mineola: Dover Publications.
Dorn, S. (2007). *Eligible but not enrolled.* Retrieved September 29, 2007, from http://www.urban.org/url.cfm?ID=411549.
Douglas-Hall, A., & Koball, H. (2006). *Basic facts about low-income children: Birth to age 18.* Retrieved August 13, 2007, from http://www.nccp.org/pub_lic05.html
Dryfoos, J. G. (1994). *Full-service schools: A revolution in health and social services for children, youth, and families.* New York: Jossey-Bass.
Dryfoos, J. G. (2002). Full-service community schools: Creating new institutions. *Phi Delta Kappan, 83*(5), 393–400.
Dryfoos, J. G. (2003). A community school in action. *Reclaiming Children and Youth, 11*(4), 203–206.
Dryfoos, J. G., & Maguire, S. (2002). *Inside: Full-service community schools.* Thousand Oaks: Corwin.
Dryfoos, J. G., Quinn, J., & Barkin, C. (2005). *Community schools in action: Lessons from a decade of practice.* Oxford: Oxford University Press.
Edwards, O. W., Mumford, V. E., Shillingford, M. A., & Serra-Roldan, R. (2007). Developmental assets: A prevention framework for students considered at risk. *Children and Schools, 29*(3), 145–153.
Fass, S., & Cauthen, N. K. (2005). *Who are America's poor children?* Retrieved August 13, 2007, from www.nccp.org/pub_cpt05b.html
Federal Interagency Forum on Child and Family Statistics. (2007). *America's children: Key national indicators of well-being 2007.* Washington, DC: U.S. Government Printing Office.

Freire, P. (1998). *Teachers as cultural workers: Letters to those who dare teach.* Boulder: Westview.

Freudenberg, N., & Ruglis, J. (2007). Reframing school dropout as a public health issue. *Preventing Chronic Disease, 4*(4), 1–11.

Geierstanger, S. P., Amaral, G., Mansour, M., & Walters, S. R. (2004). School-based health centers and academic performance: Research, challenges, and recommendations. *The Journal of School Health, 74*(9), 347–353.

Hardaway, C. R., & McLoyd, V. C. (2009). Escaping poverty and securing middle class status: How race and socioeconomic status shape mobility prospects for African Americans during the transition to adulthood. *Journal of Youth Adolescence, 38,* 242–256.

Harris, M. M., & Hoover, J. H. (2003). Overcoming adversity through community schools. *Reclaiming Children and Youth, 11*(4), 206–211.

Hilliard III, A. (2003). No mystery: Closing the achievement gap between Africans and excellence. In T. Perry, C. Steele, & A. G. Hilliard, III (Eds.), *Young gifted and Black: Promoting high achievement among African-American students* (pp. 131–166). Boston: Beacon Press.

Jago, A. G. (1982). Leadership: Perspectives in theory and research. *Management Science, 28*(3), 315–336.

Jensen, J. M., & Fraser, M. W. (2006). A risk and resilience framework for child, youth, and family policy. In J. M. Jensen & M. W. Fraser (Eds.), *Social policy for children and families: A risk and resilience perspective* (pp. 1–18). Thousand Oaks: Sage.

Jenson, J. M. (2007). Research, advocacy, and social policy: Lessons from the risk and resilience model. *Social Work Research, 31*(1), 3–5.

Keith, N. Z. (1999). Whose community schools? New discourses, old patterns. *Theory into Practice, 38*(4), 225–234.

Kitano, M. K., & Lewis, R. B. (2005). Resilience and coping: Implications for gifted children and youth. *Roeper Review, 27*(4), 200–205.

Kouzes, J. M., & Posner, B. Z. (1995). *The leadership challenge: How to keep getting extraordinary things done in organizations.* San Francisco: Jossey-Bass.

Kronick, R. F. (2005). *Full service community schools: Prevention of delinquency in students with mental illness and/or poverty.* Springfield: Charles C. Thomas.

Lee-Bayha, J., & Harrison, T. (2002). *Using school – community partnerships to bolster student learning* (Policy brief). San Francisco: WestEd.

Mandleco, B. L., & Perry, J. C. (2000). An organizational framework for conceptualizing resilience in children. *Journal of Child and Adolescent Psychiatric Nursing, 13*(3), 99–111.

Mertensmeyer, C., & Fine, M. (2000). ParentLink: A model of integration and support for parents. *Family Relations, 49*(3), 257–265.

Morris, J. E. (2008). Research, ideology, and the brown decision: Counter-narratives to the historical and contemporary representation of Black schooling. *Teachers College Record, 110*(4), 713–732.

Mykota, D. B., & Muhajarine, N. (2005). Community resilience impact on child and youth health outcomes: A neighborhood case study. *Canadian Journal of School Psychology, 20*(5), 5–20.

National Center for Children in Poverty. (2007a). *Basic facts about low-income children: Birth to age 18*. New York: Columbia University Mailman School of Public Health.

National Center for Children in Poverty. (2007b). *United States early childhood profile*. New York: Columbia University Mailman School of Public Health.

Newman, K. S., & Chen, V. T. (2007). *The missing class: Portraits of the near poor in America*. Boston: Beacon Press.

Orfield, G. (2009). *Reviving the goal of an integrated society: A 21st century challenge*. Los Angeles: The Civil Rights Project/Proyecto Derechos Civiles at UCLA.

Orthner, D. K., Jones-Sanpei, H., & Williamson, S. (2004). The resilience and strengths of low-income families. *Family Relations, 53*(2), 159–167.

Peebles-Wilkins, W. (2004). The full-service community school model. *Children and Schools, 26*(3), 131–133.

Peltzman, A., & Jerald, C. (2006). *High standards and high graduation rates*. Washington, DC: National Association of State Boards of Education.

Perry, T. (2003). Up from the parched earth: Toward a theory of African-American achievement. In T. Perry, C. Steele, & A. G. Hilliard, III (Eds.), *Young gifted and Black: Promoting high achievement among African-American students* (pp. 1–108). Boston: Beacon Press.

Planty, M., Hussar, W., Snyder, T., Provasnik, S., Kena, G., Dinkes, R., et al. (2008). *The condition of education 2008 (NCES 2008-031)*. Washington, DC: National Center for Education Statistics, Institute of Education Sciences, U.S. Department of Education.

Popper, M. (2000). The development of charismatic leaders. *Political Psychology, 21*(4), 729–744.

Reiss, S. M., Colbert, R. D., & Hebert, T. P. (2005). Understanding resilience in diverse, talented students in an urban high school. *Roeper Review, 27*(2), 110–120.

Renn, K. A. (2003). Understanding the identities of mixed-race college students through a developmental ecology lens. *Journal of College Student Development, 44*(3), 383–399.

Rice, D., & Harris, M. M. (2003). Leadership in community schools: A frame analysis. *Reclaiming Children and Youth, 11*(4), 216–211.

Richardson, J. W. (2007). Building bridges between school-based health clinics and schools. *Journal of School Health, 77*(7), 337–343.

Richardson, J. W. (2008). From risk to resilience: Promoting school – health partnerships for children. *The International Journal of Educational Reform, 17*(1), 19–36.

Rounds, K. A., & Ormsby, T. C. (2006). Health policy for children and youth. In J. M. Jensen & M. W. Fraser (Eds.), *Social Policy for Children & Families: A Risk and Resilience Perspective* (pp. 131–166). Thousand Oaks: Sage.

Sanders, M. G. (2001). The role of "Community" in comprehensive school, family, and community partnership programs. *The Elementary School Journal, 102*(1), 19–34.

Steele, C. (2003). Stereotype threat and African-American student achievement. In T. Perry, C. Steele, & A. G. Hilliard, III (Eds.), *Young gifted and Black:*

*Promoting high achievement among African-American students* (pp. 109–130). Boston: Beacon Press.

Tervalon, M., & Murray-Garcia, J. (1998). Cultural humility versus cultural competence. *Journal of Health Care for the Poor and Underserved, 9*(2), 117–125.

U.S. Department of Education – National Center for Education Statistics. (2006). *The condition of education 2006* (Vol. NCES 2006-071). Washington, DC: U.S. Government Printing Office.

United Nations Educational Scientific and Cultural Organization (UNESCO). (2002). *Focusing resources on effective school health: A FRESH start to improving the quality and equity of education.* Retrieved February 11, 2008, from http://schoolsandhealth.org/FRESH.htm.

# 3

# Pursuing Excellence in a Segregated System: Coatesville, a Town of Paradoxes

I never thought that I was lacking in anything compared to other schools that excluded us. As a matter of fact, I had great pride, love and concern from the staff.

– *Babatu Rudo aka Fred Myers,*
*words from a survey*

The conundrum of educating Black children in a society bent on devaluing them has been the lot of educators (both principals and teachers) pre-and post-*Brown*. It is beyond the scope of this volume to provide a detailed account of the nation's segregated educational system. Yet, to appreciate what JACS, as well as other similar schools, accomplished, an overview is helpful. Segregation prior to the *Brown* decision was permitted by statutes commonly referred to as Jim Crow laws. Legislated separation, more stringently enforced in the southern states, pervaded every dimension of life, to include housing, schooling, health, and retail enterprises. This is not to say that separation of Blacks and Whites was not the case in states north of the Mason – Dixon Line. The Mason – Dixon Line, though established in the late 1700s as the demarcation between slave states (Maryland and south) and free states (Pennsylvania and north), continued post-slavery to designate different legislated practices as it pertained to Blacks. That said, where segregation was *de jure* (by law) in the South or *de facto* (by practice) in the North, separation of the races was more common than not prior to the 1960s.

Although schools for Black children labored under adverse conditions (rarely funded, staffed, or resourced comparable to White schools), separation fueled the long-standing community-wide commitment to education as the great economic and social liberator. Schools in addition to churches were the institutional anchors of neighborhoods and educators were regarded as indisputable leaders. Along with high regard came a responsibility to carry oneself in a way befitting a professional and to pay for teaching supplies and books out of their meager salaries. In the absence of the solidarity and missionary-like devotion of teachers the learning prospects of students would have been abysmal (Fairclough, 2007; Walker, 1996). Furthermore, whether a predominantly Black or White community school, teachers were members of the community and as such the lines between their professional and private lives were blurred. Within the same town and city limits teachers lived, shopped, engaged in civic organizations, and attended houses of faith as the families of the children they taught. As Elsie Clapp noted, "They [the teachers] did not 'survey' the community; they belonged to it" (1971, p. ix).

Inequality of funding may have cast the building blocks for educational inequities, but they also proved fertile ground for resourcefulness and tenacity. For example, in Caswell County Training School (CCTS) in North Carolina, community members chipped in for school supplies for children even if in some cases they did not have children attending the school. Other contributed resources included the use of personal vehicles and time or skills; for example, cooks and carpenters taught young people their trades (Morris, 2008; Walker, 1996).

To understand why Blacks despite the clear disparate treatment continued to cling to educational achievement as a priority one needs to consider the power education was believed to yield. Historically, which as an aside does not appear to be the case today, teachers were second only to preachers in their stature, power, and significance in the African American community (Fairclough, 2007; Morris, 2008; Walker, 1996). Teaching was more akin to a religious calling than a professional selection (Walker, 1996). They not only conceived their roles as instructors of cognates, but also were responsible to teach children how to negotiate the majority culture. Examples of this at JACS were the etiquette classes and English teachers' demanding that "proper" English be spoken in and out of school.

Education-driven power as a benefit of group membership is also related to social and cultural capital. Cultural capital refers to the morals, skills, and practices valued by the dominant group. In other words, group membership affords certain natural exposure that can and most often does translate into advantage within a particular society. Social capital is the composite profile individuals bring to the market that would include, for example, education, economic resources, familiarity with marketplace norms, and expectations. Possession of cultural and social capital perpetuates access to opportunities to gain more cultural and social capital creating a cyclical relationship. In the United States, the role of race has been an ever-present specter, sometimes subtly and sometimes overtly controlling educational access (Richardson & Harris III, 2004). This cycle of having more (credentials) yielding more (economic and social capital) is precisely what African American educators attempted to facilitate in the nurturing of educational prowess in their students.

Fervent belief in the value of educational attainment on the part of teachers as a group in the 1940s and 1950s had its roots deeply imbedded in African American tradition and their college training. From the time it was illegal to teach a slave to read and write the quest to become educated has loomed as worth pursuing. In the words of Rev. G. M. Elliott in 1888, "Teachers you are the shapers of thought and the molders of sentiment, not of this age and of this generation alone, but of ages and generations to come. You are making history by those you teach. You are the few that are molding the masses" (Fairclough, 2007, p. 8).

When it came to securing resources from Whites, Black educators were in no position to make demands. Instead they had at their disposal only the power of persuasion and manipulation. In their own communities, their standing yielded a platform to seek support and solidarity among families and community members. Educators were living examples of how high educational attainment could solidify one's place in the middle class. Furthermore, their training equipped them to weave Eurocentric curriculum with African American tradition and history, affirming the self-esteem of students. However the danger was that by virtue of degree possession, tensions related to class differences held the potential of estranging teachers from their communities (Fairclough, 2007). Thus, educators' interest in meaningful and sustainable engagement with their community, as well as tapping into White resources could be described as a delicate balancing act.

Educators were clear about the power education could afford to their students and by inference to their entire community. They dedicated themselves to passing on acquired survival skills to students (Morris, 2004). By survival skills I refer to the higher standards of behavior and breadth of knowledge Blacks have historically needed to be offered menial positions. However, approaches set in the context of "surviving in a White man's world" alone would have undermined the confidence educators were determined to build. Ignoring explicit and implicit racially demeaning messages and practices highlighting the inferiority of African Americans have wreaked and continue to wreak havoc on the innate potential of children invoking stereotype threat, self-doubt, and feelings of dissonance with mainstream culture. So hand in hand with teaching skills valued by Eurocentric society, they interspersed culturally relevant curricula and passed on a sense of responsibility to succeed. Part of this ethos was also demonstrated in the ways teachers carried themselves when not in front of their classes.

Taking on the responsibilities associated with being a role model was an expectation ascribed to the profession. While now that might seem a less attractive and even an inappropriate expectation, teachers are role models nonetheless. No institution interacts as much with children as schools, which is why Bronfenbrenner (1976) inserted teacher – child and school – child relations in the innermost of the concentric ecological circles.

The positive side of segregated schools was their propensity to be havens from negativity. Consider CCTS, where the principal James Dillard paralleled the style and commitment of T. J. Anderson during a similar time period. Well educated with a master's degree from the University of Michigan, he was known to hold high expectations of students and teachers. In practice he frequently walked the halls, dropped into classrooms, and provided space for teachers to be creative as they implemented a rigorous academic curriculum. Like T. J., he emphasized the importance of seeing students holistically as integral to good teaching. Home visitation and after-school programming were supported as ways to keep children engaged in learning. Key to meeting the intellectual demands of students was the emphasis on professional development that would keep teachers apprised of educational trends and enhance their professional credibility (Walker, 1996).

Teachers and administrators who knew well the skills students would need in higher education and life in general dedicated themselves to be what Asa Hilliard III referred to as gap closers. Bringing all of their expertise to bear, they skillfully crafted during- and after-school interactions that supported learning and its reinforcement so that in the case of JACS students were well equipped to excel academically at the integrated high school. What does drawing comparisons between a pre-*Brown* era school portend for a school today? For one thing, segregated school populations are still a reality, as discussed in Chapter 1.

Underachievement in the face of segregated schools or those serving low-income families does not have to be the case. JACS provides a vivid example of what is possible. Taking into account the complex dilemmas facing youth and educators, several tenets have been advanced to turn lackluster educational experiences into interventions nurturing the intellectual potential of every student. Students should be exposed to caring adults; there should be safe places to learn where time is used constructively; youngsters' health should be addressed; curricula should plant the seed for not only marketable skill building but a thirst for lifelong learning; and students should be provided opportunities to help each other, thus building a sense of civic responsibility (America's Promise Alliance, 2008). Unfortunately, as young people grapple with their circumstances there is concomitantly a crisis of confidence in the one institution that has historically provided a means of upward social and economic mobility—schools.

Given the context of public schooling for so many African American students, successful practices in segregated schools are instructive. Reasons for fiscal disparities might be different, but the resulting needs of students and teachers are hauntingly similar. Let us see how Coatesville in the 1940s and 1950s might mirror some of the challenges facing school leaders today and set the context for the evolution of the Merchant Street School to the JACS.

## Coatesville: A Town of Paradoxes

*Coatesville*

*We lift our hears and eyes to see a town that is loved by men—*
*The streets cry welcome, the hills are kind and nature is our friend.*
*The houses seem to have open doors, no matter where you go,*
*Coatesville's one of those places where one finds*
*happy faces for goodwill here doth grow.*
*– Anita Anderson (A. T. Anderson, February 15, 1946)*

Coatesville was first named Bridgetown in 1714 and the name was later changed in 1787. Nestled between the Amish farmland of Lancaster and Philadelphia, the town lies in the hills along the Brandywine River. Forty miles from Philadelphia it was also a stop on the Pennsylvania Railroad main line (Washington-Williams, 2005). Coatesville was chartered in 1792 though it did not earn the status of a borough until 1867. Initially its prominence was grounded in its location, midway between Philadelphia—the nation's capital at the time—and Lancaster—the second largest urban center in the Commonwealth. Shortly thereafter, notoriety was tied to Lukens Steel Mill, at one point home to the largest steel plates in the world (Woodward, 2005).

Moses Coates, for whom the town is named, was the first postmaster, and based on his Quaker faith was vested in freedom for slaves and the anti-liquor movement (Woodward, 2005). The Society of Friends (Quakers) is a Christian religion opposed to war, violence, and slavery. Members along with other religious sects (such as Mennonites) commonly hosted Underground Railroad "stations" on their properties. The Underground Railroad was a secretive network of individuals who defied the law by aiding Blacks escaping slavery (Hopkins & Smith, 1994). While in the "stations," slaves would be fed and directed to the next safe haven with the route extending from the southern U.S. states to Canada. Coates's home in East Caln served as a station of the Underground Railroad, demonstrating his commitment to the abolition of slavery.

Coatesville's history is peppered with the waves of immigrants of Scotch-Irish and Italian descent, grandchildren of slaves, and ecumenical groups that settled in the region. Quakers represented the largest religious group, followed by Methodists and Presbyterians. The role of Negroes in the establishment of Coatesville is unclear; however, given their continuous passage through Underground Railroad stations, and the location of the town north of the Mason – Dixon Line, it is reasonable to assume that after the abolition of slavery they might be apt to stay in a supportive Quaker settlement. Evidence of Quaker sentiments can be found in Coates family anecdotes, which reveal their concern for their "dark-skinned brethren" (T. J. Anderson, 1950b; Woodward, 2005).

African Americans fought at both Brandywine and Valley Forge and subsequently won their freedom. This freedom was afforded whether they fought for the British or the American colonies. In 1780, Pennsylvania formally ended slavery with what was called a gradual emancipation. No new slaves could be bought or sold and no slaves born after 1780 could be a slave after the age of 28. Given the abolitionist tenor in the Commonwealth, it is no surprise that Black communities were established as early as the 1840s. Two such hamlets were Hayti (pronounced

"Hay-tie") and a section of the city called East End (Hopkins & Smith, 1994). The history of Coatesville is in the telling a very American saga. Early colonial governmental struggles in Pennsylvania pitted the Quaker majority, who were pacifist and did not want to go to war against the British, against the Scotch-Irish population who wanted to declare war. The most famous of the early patriots was John Gleming, who served in the 1776 convention that framed the first constitution and was a member of the Pennsylvania General Assembly. Scotch-Irish settlers fleeing the religious and economic discrimination they faced in Northern Ireland were welcomed by the Quakers, and by 1729 it is estimated that there were over 6000 Scotch-Irish residing in William Penn's colony who primarily practiced their faith as Presbyterians. They established the Upper Octorara Presbyterian Church (later pastored by James Brown, a friend to T. J. Anderson and JACS).

It is no surprise that the pastor of Upper Octorara Presbyterian Church would be at the forefront of social change and more specifically integrating schools. The church, since its inception in 1720, found itself entrenched in political and social movements. Rev. William Foster preached impassioned sermons to arouse patriotism and the British sought to capture him and burn down the church, though unsuccessfully. He along with his brother-in-law Rev. John Carmichael were good friends of Benjamin Franklin (Upper Octorara Presbyterian Church, 2008; Woodward, 2005).

The Lukens Steel Mill was founded by Isaac Pennock in 1810 and was run after his death by his daughter Rebecca Lukens. She was a Quaker widow at the time, who for 50 years directed the company, quite an unusual responsibility for a woman at that time in history. Later, her son-in-law Dr. Charles Huston was credited with recasting the iron works company into a modern steel enterprise. At one point, Coatesville boasted possession of the largest steel plates in the world (Woodward, 2005).

One of the first churches for Blacks was erected in 1820. Originally named Union American Methodist Episcopal Church, the edifice was renamed in 1869 in honor of Rev. William Hutchinson. The church was located in what became the thriving Black section of town, located at Sixth and Merchant Avenue, and was later moved to Eighth and Chestnut Ave in 1908, where it remains and stands today under the name of Hutchinson Memorial United African Methodist Episcopal Church (Woodward, 2005).

By 1860, of the 633 residents of Coatesville, 49 were free Negroes born above the Mason – Dixon line. Not unlike other colonies, though the first Blacks came to Pennsylvania as slaves, they were as likely to work as sail makers, bakers, and carpenters as farmhands or house servants.

The demise of the slave labor in Pennsylvania was predictably prompted by the resistance advanced by the Quakers.

Coatesville's racial intolerance notoriety came from the brutal lynching of Zack Walker in 1911. Walker was a young Black Virginian who had moved to Coatesville for work on the steel assembly line. After a night of drinking Zack was walking home and he fired his gun in the air. It was not uncommon for steel workers Black or White to carry guns at the time. Edgar Rice, a former policeman, attempted to arrest Walker; a struggle ensued and Rice was killed. Search parties were assembled to find "the Black fiend." A terrified Zack climbed up a tree and attempted suicide unsuccessfully. He was taken to Coatesville Hospital, only to be forcibly taken a few days later by a mob of over 1000 White men. Grisly accounts of the lynching number the crowd at 4000, who cheered and chanted for Walker to be burned. Though apparently escaping the flames three times, he was eventually burned to death (Hopkins & Smith, 1994; Washington-Williams, 2005). Shortly thereafter, NAACP called for an investigation and established three local chapters in Coatesville, Harrisburg, and Philadelphia. There was an uneasy truce of sorts between Blacks and Whites after the nationally publicized lynching because of the need for workers in iron and steel mills.

Though apparently less restrictive than cities below the Mason – Dixon line, segregation was practiced in Coatesville in housing, retail, schooling, and health care. Nevertheless, despite segregation and the Zack Walker lynching, Coatesville remained attractive to many southern Blacks because of the good pay and opportunity to live outside of Jim Crow segregation restrictions imposed 20 miles south in Maryland.

The great migration of Blacks to Coatesville began in earnest after 1915. Companies like Lukens Steel Mill were in such a need of workers that they convinced railroads to provide free transportation to migrants; the employers would deduct from subsequent wages the cost of the train ticket (Hopkins & Smith, 1994). Blacks were critical to the economic life of the town and at the same time were actively excluded from decision making and upper echelons of town life. Despite the need for workers in the mills, many Blacks did not find the equitable opportunities they anticipated, partially due to exclusion from the unions, relegation to the jobs that were most dangerous, and lowest wages on the wage scale. Companies did not go out of their way to provide housing to Black migrants, so they often ended up in already overcrowded neighborhoods. However, this relegation to second-class citizenship had some unanticipated consequences (Woodward, 2005).

Segregating Blacks to certain neighborhoods and parts of town provided opportunities for entrepreneurship. Cottage industries such as barbershops, funeral homes, caterers, insurance brokers, and stores created a city within a city. Additionally, very near to Coatesville were two historically Black colleges (HBCUs) (Cheney State and Lincoln University) that began as normal schools and evolved into full-service universities providing an educated workforce prepared to teach, and pursue medicine, the ministry, and other occupations (Hopkins & Smith, 1994). Two aforementioned churches specifically dedicated to Black congregants were noteworthy as community anchors: the United African Methodist Episcopal (later named Hutchinson Memorial UAME) Church and St. Paul's African Methodist Episcopal Church (Woodward, 2005).

In 1920, there was a second wave of immigrants—most of them from Italy, Russia, and Poland—boosting the town's population to 14,500. Enamored with the quality of life Coatesville afforded, the Board of Trade in the late nineteenth century wrote that Coatesville was an ideal town to settle:

> Its scenery is unsurpassed anywhere.... Good churches, good schools, good markets, elegant streets, handsome and tasty residences in perfect repair and order, flourishing societies of all sorts, an extensive public library... it is indeed the ideal town for all those who like the country and desire a happy place in which to live and do business. (Woodward, 2005, p. 33)

In the 1920s and 1930s the racial tensions were fueled again by an infiltration of the Ku Klux Klan (a White supremacy organization). Then in 1938 a near repeat of the Zack Walker incident occurred. A young White woman was assaulted and raped. When she spoke to police, she told them the perpetrator had been a Negro. Nearly 500 White men formed a posse to root out the culprit. The suspect was a former high school football player who worked at Lukens; he was taken into custody. Talk of a lynching prompted not only a mob of Whites, but this time a large contingent of armed Blacks to descend on the jail. It was only when the police chief reminded the crowd of what had happened to Zack Walker and how the event had been "the shame of the nation" that the mere mention of the name calmed the crowd down. As it turned out, within a week a White man confessed to the assault and was arrested (Washington-Williams, 2005; Woodward, 2005). One again an uneasy truce was struck.

Lukens Steel Mill and the Veterans Administration Hospital were the principal employers between 1943 and 1956 and at least a decade after that period. The mill was a major producer of World War II (WWII) supplies, which given the founder's Quaker affiliation was quite ironic. In 1944, the nation's first Liberty ship named after a woman industrialist was commissioned and named the SS *Rebecca Lukens* and christened by her great-granddaughter, Ruth Huston, in honor of the mill's support during WWII. As additional evidence of Lukens's high regard in the war effort, Eleanor Roosevelt sent notes not only to Ms. Huston but also to the women employees of Lukens, who kept the plant running in the absence of their husbands, sons, and fathers (Woodward, 2005).

Coatesville by this time was the industrial heartbeat of Chester County, principally as a function of the expansion of Lukens Steel Company and other companies such as DuPont, Wyeth, Inc., and Grocery Store Products Company (Chester County Historical Society Archives, January 17, 1941, June 15, 1950). Coatesville had also become the site of a large veterans' administration hospital. Important to note is the suburb effect that began to take place, which was most likely a function of the availability of good-paying work. The dispersal of the population created suburbs in border townships like Caln (whose population increased by about 2000 between 1930 and 1940) and South Coatesville (The Chester County Branch of the Pennsylvania Economy League Inc., 1950). Needless to say, a burgeoning numbers of residents precipitated an increased demand for housing.

In the early 1940s the population boom prompted the construction of housing units in a 3-mile radius of Coatesville. One hundred and fifty housing units for Whites in an area called Brandywine Homes and 100 units for Blacks in Carvers Court were two examples. Other developments were in Rock Run, Westwood in Hayti (for Negroes), and Rainbow Village in Pomeroy Heights. Unfortunately, the buildings did not keep pace with the needs of particularly single men who moved to the city for work (Chester County Historical Society Archives, August 25, 1951, June 16, 1950; The Chester County Branch of the Pennsylvania Economy League Inc., 1950).

A more appropriate assessment of Coatesville's population at the time would include the surrounding 11 townships (Caln, South Coatesville, East Fallowfield, Honeybrook [Township and Borough], Highland, Sadsbury, Valley, West Bradford, West Brandywine, and West Caln). Inclusion of these residential areas brought the population of the business and industry—the trading area of Coatesville—to approximately 50,000 in 1950, more than a third of the entire county's population (Chester County Historical Society Archives, August 25, 1951, June 16, 1950; The Chester County Branch of the Pennsylvania Economy League Inc., 1950).

Blacks could shop at some establishments, such as Santee's Drug Store, but were not permitted to sit at their soda fountain counter. Blacks could not swim in the public city swimming pool, but could go to French Creek, which flowed through a nearby state park. By 1948, a swimming pool was opened on Ninth Avenue in the heart of a Black residential district, East End. Two movie theaters in town segregated Blacks to the "colored balcony" or the rear of the theater. Though technically all residents could seek medical attention at Coatesville Hospital, treatment of Black patients was so poor that Dr. Whittier C. Atkinson opened a hospital in the East End in 1937 after struggling to ensure his patients' treatment since his arrival in 1927 (Washington-Williams, 2005; Woodward, 2005).

## Coatesville Interracial Committee

The Coatesville Interracial Committee was established in 1943 at the behest of Governor Martin who charged Mayor Luebert with its creation. As noted in the local newspaper, the organization was to be part of a larger nationwide movement to "foster better relations between Whites and Negroes" (Concern, November 18, 1946c). Power in policy circles is discussed in terms of formal and informal networks. While it does not appear the committee was afforded any formal authority, the fact remains that meetings were held in City Hall and the minutes saved in the public records are similar to the City Council minutes, alluding to the "bully pulpit" and informal power of the group.

In the year T. J. Anderson was hired to return to Coatesville (1943), he and several other community leaders were asked to serve on the Interracial Committee. At the first official convening, Rev. James Brown, Pastor of the Upper Octorara Presbyterian Church, was elected Chair with other elected posts held by Leon Prout and Mrs. Sadie Watters Taylor. Additional members of the committee were Walter Greenwood, Rev. Arthur Younger, T. J. Anderson, William Muthard, and John T. Cuff (Chester County Historical Society Archives, October 19, 1943). As an aside, many of these committee members became lifelong friends.

The committee's two affirmative objectives were (1) to establish programming that would "minimize or remove some of the sources of racial friction, and (2) promote interracial goodwill." The purpose of specific panels and cultural exchanges was to cultivate contact between persons of different races; engage in daily living activities together; and, for those interested persons, inform them of major aspects of race relations. One of the first decisions of the committee was to commission a report on the state of interracial relations in Coatesville (Chester County Historical Society Archives, October 19, 1943).

Ernest Atwell was hired to analyze the nature of race relations in housing, education, and retail. At the time he was director of the Bureau for Colored Work of the National Recreational Association (the precursor to the national Health, Physical Education and Recreation Association [HPER]), was a colleague of Booker T. Washington, and had worked in the

Herbert Hoover administration. Based on his observations he suggested, "the nucleus of the organized recreation program of the Adams school would be easily transferable and should the center project become a reality" would represent a lifetime investment. On the heels of his comments, the Interracial Committee officials met with Coatesville Chamber of Commerce and the East End Recreation Association to propose a replication of the JACS model (Chester County Historical Society Archives, June 16, 1944; Concern, December 15, 1945a). Directors of industry were also interviewed to encourage an elimination of segregation in town theaters and at workplaces (Woodward, 2005).

As an aside, reconstruction of JACS programming and community influence in this volume is heavily supported through archived articles from the local news paper, *The Coatesville Record*. Though now out of print, intensive news coverage of JACS may have been at least in part influenced by findings of the Interracial Committee reports that urged the publisher of the paper to print news helpful in nurturing "wholesome community attitudes" (Chester County Historical Society Archives, November 15, 1943). Strategies that informed the group well into the 1950s were the need to gather community sentiments about the challenges and secure a commitment to effectively use the local newspaper (*The Coatesville Record*) to advance the work of the committee.

The Interracial Committee also hosted integrated cultural and scholarly discourse, held public meetings, and eventually published their own periodical of interracial progress called *Concern*. It is interesting to note that in the context of discussing separate schools from a philosophical standpoint, Whites and Blacks in Coatesville referred to the practice as parallelism. However, in advocacy documents, such as those published by the Interracial Committee, the term segregation was used.

In its maiden printing, the purpose of the publication *Concern* was stated as follows: "The American way of living together demands that each recognize in the other the worth of individual personality.... It is our 'concern' that we all be bridge-builders in a day that has produced great chasms in the earth" (*Concern*, December 15, 1945b). Regular columns included poetry (one of the contributors being Melvin B. Tolson, the professor portrayed in the recent movie *The Great Debaters*), Voltaire, Anita Anderson, and many others. Editorial columns, book reviews, and articles reported local forums and advocacy targets also populated pages. A regular column, "On the Local Interracial Front," reported advances in integration, peaceful resistance to segregation, and individuals or groups who advanced the cause of racial understanding.

Some articles were designed to inform well-meaning Whites how to best interact with Negroes, such as "On Interracial Etiquette," which included suggestions such as "A sentimental tale about one's 'old black mammy' will prevent or destroy rapport with a Negro audience"; another example was "The Negro population has upper- middle- and lower-class distinctions which closely approximate those of the White population" (*Concern*, February 15, 1946a). Assuming a helpful tone was designed to embrace persons of any background interested in pursuing social justice.

Entries in "On the Interracial Front" included kudos for cracks in the segregation ceiling of public life such as former JACS students, who while matriculating in the senior high school broke color barriers. For example, readers were informed when Theodore Moody was elected president of the integrated sophomore class of 1946 or when Dr. Stokes purchased a home in a previously segregated neighborhood (*Concern*, February 15, 1946b).

Panel discussions hosted by the committee focused on educational, political, economic, social, and psychological topics and were open to the general public (Coatesville Interracial Committee, November 15, 1943). One November evening in 1945, attendees discussed the "problem of teaching literature which portrays unfavorable characters as representative of whole races to pupils who are susceptible to impression without the knowledge or maturity to counteract unbalanced characterization." It was decided at the conclusion of this specific event that a paper would be commissioned to deconstruct the aforementioned challenge and make curricular recommendations to combat it (Chester County Historical Society Archives, November 20, 1945).

Also fundamental to the cross-cultural participation on the committee was a commitment to religious tolerance. Rabbis, and members of the Quaker, Presbyterian, Methodist, Baptist, and Seventh Day Adventist faith communities were active members of the committee. Various professional affiliations were also represented, such as physicians, and business persons. Cultural events were regularly hosted or cohosted with JACS. Among the many guests of the committee were the child pianist prodigy Philippa Duke Schuyler (Chester County Historical Society Archives, February 21, 1944) and famed singer Roland Hayes (Chester County Historical Society Archives, December 31, 1945). Those of an advocacy bent included Dr. Wallace Gallup, who at the time was the regional director of the National Conference of Christians and Jews (Chester County Historical Society Archives, January 21, 1947), Dr. Morris Wolf, professor of government at Girard College in Philadelphia and member of the World Affairs Council of the United Nations (Chester County Historical Society Archives, May 16, 1950), and Dr. Howard Thurman, the renowned theologian (Coatesville Interracial Committee, November 15, 1943).

By 1943, the paradoxes of life in Coatesville were evident in numerous quadrants. First the city's Quaker roots make the demonstrative segregation of Blacks out of sync with the founders' sensibilities. Though north of the Mason – Dixon Line, segregated housing and retail practices paralleled those enforced in the South, yet there was only partial devotion to segregation in terms of education. Even in the establishment of school philosophies that were a matter of public record, there was segregation except in Grades 10–12 (Chester County Historical Society Archives, 1954).

Tension between generally accepted practice and social justice seemed ever in the consciousness of some Whites and Blacks (Washington-Williams, 2005; Woodward, 2005). It is in the heart of these paradoxes that JACS was destined to evolve.

### James Adams Community School

*Yes, I am your school hold your heads high*
*Because of you I shall never die*
*You are my hope of days to be*
*Your school spirit helps keep me free.*
*The day will come when in your place*
*Children of every creed and race*
*Will stand, for I was meant to be*
*A symbol of Democracy.*
   – an excerpt from *I Am Your School*
   (Chester County Historical Society
   Archives, June 1954)

Merchant School was founded in 1906, and was located in the East End of town in the heart of the African American residential neighborhood (D'Angelo, 2003; Monroe-Waters, 1942). It began as a three-story building with two rooms on each floor for first through fifth grades. For a cost of $14,932 the building was erected and opened its doors on December 12, 1906, to 175 students with five teachers employed by the school district. The following year sixth grade was added (T. J. Anderson, 1950a; Holton, 1951).

Beginning in 1907, grades were added until it became a school serving kindergarten through ninth grade after which students attended the integrated high school for tenth, eleventh, and twelfth grades. An addition expanded the population capacity of the school in 1924, which also allowed for the hiring of additional teachers. During the same school year, vocational courses in carpentry and painting for males and home economics for girls were added to provide a degree of curricular parity with the White schools of the town (T. J. Anderson, 1950a).

Until 1920, Merchant School only taught classes up to the sixth grade after which students had to travel half a mile away to a desegregated school (D'Angelo, 2003). Disputing the logic of this practice a report issued by then Superintendent H. R. Vanderslice touted the desegregation of middle school grades (sixth to eighth) as an abysmal failure in terms of promoting graduation rates. For example, between 1878 and 1923, Coatesville High School issued only 16 diplomas to Blacks (Chester County Historical Society Archives, December 1950). Opposing the way "Negro" students were served in the desegregated middle grades in no way made Vanderslice a segregation proponent. His concern was that students were not graduating and deemed that fact problematic. However, limitations imposed by the school board prohibited equitable support for Black students in White schools and, as a result, Vanderslice's strategy shifted to ensure the most parity possible in finances, facilities, faculty credentials, and curricula in segregated schools.

Shortly thereafter in 1924, the school board added extra rooms, a gymnasium, two shower rooms, four additional classrooms, a kitchen, and stage to Merchant's edifice. This marked a new era for Merchant School in other ways as well. Superintendent Vanderslice actively lobbied for the hiring of additional credentialed teachers, authorized the expansion of the curriculum, and actively recruited T. J. Anderson, a recent graduate from Howard University, as principal. Upon the opening of the expanded facility the name of the school was changed to honor a Coatesville resident, James M. Adams (D'Angelo, 2003; Monroe-Waters, 1942).

James Morgan Adams, a local native, was an industrious entrepreneur owning his own moving and hauling business and considerable real estate. He was an active member of the local United African Methodist Episcopal Church, known also as Hutchinson UAME. While any of these accomplishments is noteworthy, particularly given his birth in 1842, the distinction that earned the naming of the school after him was his being the only Negro who had until that time ever served on the City Council of Coatesville (T. J. Anderson, 1950a).

In the summer of 1925, concerned about disintegrating public relations between James Adams School (as it was called then) and the community, Vanderslice called Howard University in search of a dynamic individual who could be trained to be a principal. The Dean of the college replied, "I have just the person you seek" and recommended T. J. Anderson. An interview followed and a few days later T. J. was in Coatesville. Shortly after arriving, an agent of Milton Bradley Company of Philadelphia in a meeting with T. J. warned that Coatesville was "the greatest hell hole in the state of Pennsylvania.... I've traveled all over this state and I'm telling you there is nothing like this school in Pennsylvania. White people are afraid to come into this section of town after sunset" (T. J. Anderson, 1950a, p. 9). The comments only confirmed the concerns held by Superintendent Vanderslice, which T. J. attributed to a quarter century of mutual distrust, overcrowded housing, and isolation (T. J. Anderson, 1950a).

Given the enormity of the task ahead, during his first and most notably his second tenure as principal, Anderson attributes four factors to the change in the school – community climate: highly trained and resourceful faculty; programs that increased parent and citizen morale; progressive administrative leadership to include the school board; and the silent partnership of the local Young Men's Christian Association (YMCA) (T. J. Anderson, 1950a).

By our standards today, there were relatively few principals by contemporary standards over the school's lifespan. Table 3.1 provides the names of principals in chronological order.

In 1929, T. J. left Coatesville with his family and ended up after a short stint in Indianapolis, then on the faculty of Howard University. Mr. Brooks and Ms. Sadie Waters carried on the tradition of engaging the community with some of the programs initiated by the Andersons, such as the oratorical contests and evening adult literacy classes.

Philosophical tenets serving as JACS's compass for professional and extracurricular activities always stressed social responsibility. For example, prior to Anderson's return, Principal Sadie Monroe-Waters (1942) advanced the school motto, "lifting as we climb." Later CIHU, the acronym for "Can I Help You," became the clarion call, which will be discussed more in depth later.

Table 3.1  James Adams Principals

| Principals | Dates |
|---|---|
| Henry Bantum | 1906–1907 |
| Rev. Rylond | 1907–1914 |
| J. W. O. Garrett | |
| Lorenzo Morris | 1914–1918 |
| John T. Cuff | 1918–1919 |
| Harrold Denny | 1919–1925 |
| T. J. Anderson | 1925–1929 |
| Walter Brooks | 1929–1941 |
| Sadie B. Waters | 1941–1943 |
| T. J. Anderson | 1943–1956 |
| *Elizabeth W. Gibson | 1956–1963 |

*During this period JACS was being phased out by virtue of the *Brown v. Board of Education* decision.
Source: *Adams Gazette*, June 1950.

On October 2, 1940, a serious fire forced students and teachers from the building and completely destroyed the library and several of the first grade classrooms. No one was hurt, which was remarkable given the 16 teachers and 480 children who were in attendance at the time (D'Angelo, 2003). During the renovation, classes were held after traditional school hours in the White schools in town (Cuff, 2007; D'Angelo, 2003). Growth of the school population, however, meant that even after returning to the renovated structure in February of 1941, a new addition was needed. Said addition was dedicated on April 9, 1942.

In the summer of 1943 at the behest of the school board the Andersons returned to Coatesville to resume leadership of James Adams. T. J.'s goal was to address community needs inadequately met by existing agencies and move beyond a parallel system to an integrated school experience (T. J. Anderson, 1950c). Over time the role of the school community in strategic advocacy was acknowledged and it was during this period (1943–1956) that "community" was added to the official name of the school. Articulated in the school philosophy was a commitment to

> [s]hape an environment where the clientele will develop more fully and freely in an atmosphere of security and affection while, at the same time it interprets to the total population the aims and aspirations of the minority group to become first-class citizens is the *sine quanon* of the segregated junior high school. (Chester County Historical Society Archives, June 1952, p. 2)

Clearly Mr. and Mrs. Anderson (who were consistently remembered as an inseparable leadership team) viewed the school as a community problem-solving institution (Britton & Britton, 1970).

It is worth noting that though segregated, James Adams during the second appointment of T. J. Anderson bested White schools in some respects. In 1946 JACS was the first public school in Coatesville to be granted a charter by the National Junior Honor Society (D'Angelo, 2003). JACS also had more teachers with master's degrees on staff than any other public school in Coatesville for several years during the Anderson tenure.

To give a snapshot of the distribution of students in Coatesville let us consider the school year 1948–9. Total city school enrollments of 3179 were taught in six schools by 151 teachers. Three (White) elementary schools (Benner, Columbia, and Craig Ridgeway) ranging from kindergarten through sixth grade served 1035 students. Gordon Junior High (White), home to seventh through ninth grades, hosted 714, and the integrated Scott Senior High School had a student body of 945. JACS taught grades kindergarten through ninth grade for Blacks and had a student body of 485

(The Chester County Branch of the Pennsylvania Economy League Inc., 1950).

Nirvana, JACS was not. The physical plant seemed perpetually in need of repair and inadequate to service the ever-growing student body. Staff support was insufficient given the size of the student body and comparable staffing at the White schools and teachers were acknowledged as overworked. Equipment was worn, financial resources always seemed to fall short of meeting needs, and, because of all these factors, curricular offerings had historically been limited (The Chester County Branch of the Pennsylvania Economy League Inc., 1950). Cognizant that parallelism impacts the psyche of children, JACS faculty and staff did what they could to instill a counter-narrative of possibility. They reminded students they were citizens of the world and reinforced this precept through current event panel discussions at graduation ceremonies, which included topics of international/global consequence. Speakers were regularly present to not only provide inspiration, but also demonstrate by virtue of their accomplishments what was possible if students applied themselves.

Rather than dwelling on inequities, efforts, particularly in the second Anderson era, were directed toward doing as much as possible with what was available, improving on old ideas, capitalizing on the best ideas of staff, pupils, parents, faculty, and friends, seeking cooperation from others in support of JACS goals, and instilling in pupils a sense of responsibility beyond themselves embodied in the CIHU school model (Chester County Historical Society Archives, June 1952). That is not to say that resisting the status quo was sacrificed. By virtue of invited guest speakers, articles in *The Adams Gazette*, *The Coatesville Record*, and *Concern* (the monthly Interracial Committee publication), and presence at public meetings, the political positions of the African American community were well represented. As one example, though his affiliation with the Interracial Committee and its publication was external to his school-based duties, Mr. Anderson used the meetings and articles in *Concern* to highlight the community programming at JACS and secure resources to underwrite the cost of said programming.

It would be remiss of me to ignore the parallelism evident in the school-sponsored sports programs during our focal period. Coatesville historically fancied itself a sports town. Athletic acumen of JACS students was almost legendary. However, the segregation of Blacks was not apparent in the record books. For example, in 1952, Rod Perry and Art Pollard broke city and county records in track and field for three straight years. There were no footnotes that though running in their earlier years for the Gordon Junior High School team, they actually matriculated at JACS. Young men who wanted to play football could do so under the supervision of their JACS coach Mr. William (Pop) Ransom, on the Gordon Junior High team. Track teams were similarly structured until 1950, when James Adams broke from tradition and supported their own track team identity.

Much text will be devoted to the programming and organization of the school during Anderson's second appointment (1943–56). For now, it is sufficient to know that full-service school status based on current definitions was achieved during this tenure. Parting words from the graduation class of JACS in 1954 best express the cumulative manifestation of this highly functioning community school. In the graduation edition of *The Adam Gazette* students of the graduation panel crafted an article entitled "I Am the Adams Community School." Excerpts from it demonstrate how attendance infiltrated the ways of viewing the world:

> I am the Adams Community School. Born a stepchild in 1906, I have grown to become the symbol of family security and affection...
>
> Equality of opportunity is our family motto. Segregation and I are deadly enemies. I fight to make Coatesville a better place to live in. The doors of my home are open day and night. Whosoever will may come and drink at the fountain of knowledge freely. My services encompass the old as well as the young. Total education is served at my table. Can I Help You is the vitamin which keeps me alive. My children learn to do by doing their work as well as their playing...
>
> I am T. J. Anderson the Principal of the Adams school. Twenty-eight years ago I dreamed a dream about a school which would serve as a school by day and a community center by night. I converted my dream into a reality!
>
> I am the parents of the pupils in the Adams Community School.... Grant me the power to guard their health, education, recreation, and spiritual development, and make them strong in body, mind, and spirit.
>
> I am the Adams teachers. My work goes home with me and on into the night. I take it with me at vacation time, and to church—it shares my dreams and my waking hours.
>
> I am the Adams pupil. The Adams community school is built around me. I am the student. It was for me and many others like me that they took the natural resources of the earth and cut, fixed, and shaped them into a school.... I am the one who is expected to hold his head high and help push my race forward.
>
> I am the Cihu Spirit, the motto of the Adams School. Cihu means, "Can I Help You" I give purpose to learning. I am the very heart of the school. I radiate joy in giving to hospitals, community services and national charities. I am the concern which you develop for your classmates, your

neighbors and for mankind. (Chester County Historical Society Archives, June 1954; Graduation Panel Class 1954, June 1954)

– Excerpts from *I Am Your School* (Chester County Historical Society Archives, June 1954; Graduation Panel Class 1954, June 1954).

James Adams was a community school not only in name, but also in the ways its life became synonymous with the life of the Black community in the East End of Coatesville. The evolution of James Adams into James Adams Community School (JACS) would not have been possible in the absence of a strong leader with leadership acumen. Without exception, Mr. and Mrs. Anderson have been given credit for the school's transformation into a full-service community school. The following chapter begins a deconstruction of the HECS model starting with the leader and leadership characteristics demonstrated by the Andersons that supported the creation and sustained community school status of James Adams.

## References

Adams Gazette. (June 1950). *Adams through the years* (Vol. 25). Coatesville, PA: Adams Community School.
America's Promise Alliance. (2008). *Every child every promise: Turning failure into action*. Retrieved February 2, 2009, from http://www.americaspromise.org/uploadedFiles/AmericasPromiseAlliance/Every_Child_Every_Promise/ECEP_Reports_-_JPEG/ECEP%20-%20Full%20Report.pdf.
Anderson, A. T. (February 15, 1946). *Coatesville* (Vol. 1). Coatesville, PA: Concern, Chester County Historical Society Archives.
Anderson, T. J. (1950a). Adams through the years. *Adams Gazette, 25*(2), 9.
Anderson, T. J. (1950b). Early history of Coatesville. *Adams Gazette, 25*(2), 5,7.
Anderson, T. J. (1950c). The Negro in Coatesville. *Adams Gazette, 25*(2), 5–6.
Britton, J. O., & Britton, J. H. (1970). Schools serving the total family and community. *The Family Coordinator, 19*(4), 308–316.
Bronfenbrenner, U. (1976). The experimental ecology of education. *Educational Researcher, 5*(9), 5–15.
The Chester County Branch of the Pennsylvania Economy League Inc. (1950). *A study of the school plant facilities in the Coatesville area of Chester county Pennsylvania*. West Chester: The Chester County Branch of the Pennsylvania Economy League Inc.
Chester County Historical Society Archives. (January 17, 1941). *Chester county population*. West Chester: Daily Local News.
Chester County Historical Society Archives. (October 19, 1943). *The Coatesville Interracial Committee*. Coatesville, PA: Coatesville Record.
Chester County Historical Society Archives. (November 15, 1943). *Coatesville Interracial Committee: 1943–1944 program proposal*. Coatesville, PA: Coatesville Interracial Committee.

Chester County Historical Society Archives. (February 21, 1944). *Children's cultural hour*. Coatesville, PA: Coatesville Record.
Chester County Historical Society Archives. (June 16, 1944). *Attwell's assessment of Coatesville recreation facilities for Negroes*. Coatesville, PA: Coatesville Record.
Chester County Historical Society Archives. (November 20, 1945). *Greater Coatesville Interracial Committee November meeting*. Coatesville, PA: Coatesville Record.
Chester County Historical Society Archives. (December 31, 1945). *A song recital by Roland Hayes*. Coatesville, PA: Coatesville Record.
Chester County Historical Society Archives. (January 21, 1947). *January meeting of the Coatesville Interracial Committee*. Coatesville, PA: Coatesville Record.
Chester County Historical Society Archives. (June 15, 1950). *West Chester growing*. West Chester: Daily Local News.
Chester County Historical Society Archives. (June 16, 1950). *Coatesville—south Coatesville most populous in county*. Coatesville, PA: Coatesville Record.
Chester County Historical Society Archives. (May 16, 1950). *Greater Coatesville Interracial Committee*. Coatesville, PA: Coatesville Record.
Chester County Historical Society Archives. (December 1950). *The 1923 report by superintendent H.R. Vanderslice*. Coatesville, PA: Adams Gazette.
Chester County Historical Society Archives. (August 25, 1951). *How many people live in Coatesville*. Coatesville, PA: Coatesville Record.
Chester County Historical Society Archives. (June 1952). *Objectives and philosophy of the Adams school*. Coatesville, PA: Adams Gazette.
Chester County Historical Society Archives. (June 1954). *I am your school*. Coatesville, PA: Adams Gazette.
Chester County Historical Society Archives. (1954). *Teachers' handbook*. Coatesville, PA: Coatesville City School District.
Clapp, E. R. (1971). *Community schools in action*. New York: Arno Press & The New York Times.
Coatesville Interracial Committee. (November 15, 1943). *Proposed program Coatesville Interracial Committee 1943–1944*. Unpublished manuscript, Chester County Historical Society Archives, West Chester.
Concern. (December 15, 1945a). *Highlights of Interracial Committee's program* (Vol. 1). Coatesville, PA: Chester County Historical Society Archives.
Concern. (December 15, 1945b). *Our concern* (Vol. 1). Coatesville, PA: Chester County Historical Society Archives.
Concern. (February 15, 1946a). *On interracial etiquette* (Vol. 1). Coatesville, PA: Chester County Historical Society Archives.
Concern. (February 15, 1946b). *On the interracial front* (Vol. 1). Coatesville, PA: Chester County Historical Society Archives.
Concern. (November 18, 1946c). *Rev. James H. Brown reconsiders resignation* (Vol. 2). Coatesville, PA: Chester County Historical Society Archives.
Cuff, B. (2007). *Interview*. Coatesville, PA: J. W. Richardson.
D'Angelo, J. (2003). *Remembering James M. Adams*. Coatesville, PA: Chester County Historical Society Archives, Coatesville Ledger.

Fairclough, A. (2007). *A class of their own: Black teachers in the segregated South.* Cambridge, MA: The Belknap Press of Harvard University Press.

Graduation Panel Class 1954. (June 1954). *I am the Adams community school.* Coatesville, PA: Chester County Historical Society Archives, Adams Gazette.

Holton, L. A. (1951). The James Adams School. (Supplied by Chester County Historical Society from Holton to Coatesville School Board on April 12, 1951.)

Hopkins, L., & Smith, E. L. (1994). *The African Americans in Pennsylvania* (Vol. Pamphlet 6). Philadelphia: The Pennsylvania Historical and Museum Commission.

Monroe-Waters, S. B. (1942). *James Adams School guide book–1942.* Coatesville, PA: James Adams School.

Morris, J. E. (2004). Can anything good come from Nazareth? Race, class, and African American schooling and community in the urban South and Midwest. *American Educational Research Journal, 41*(1), 69–112.

Morris, J. E. (2008). Research, ideology, and the Brown decision: Counternarratives to the historical and contemporary representation of Black schooling. *Teachers College Record, 110*(4), 713–732.

Richardson, J. W., & Harris, J. J., III (2004). Brown and its higher educational legacy: HBCUs and their 'separate and unequal' standing. *The Journal of Negro Education, 73*(3), 365–378 Upper Octorara Presbyterian Church. (2008). *Our History.* Retrieved July 29, 2008, from http://www.uopchurch.org/Pages/history.html.

Walker, V. S. (1996). *Their highest potential: An African American school community in the segregated south.* Chapel Hill: The University of North Carolina Press.

Washington-Williams, E. M. (2005). *Dear senator: A memoir by the daughter of Strom Thurmond.* New York: ReganBooks.

Woodward, W. C. (2005). *Coatesville yesteryears.* West Chester: Chester County Historical Society.

# 4

# The Cornerstones: The Blurred Personal and Professional Lives of T. J. and Anita Anderson

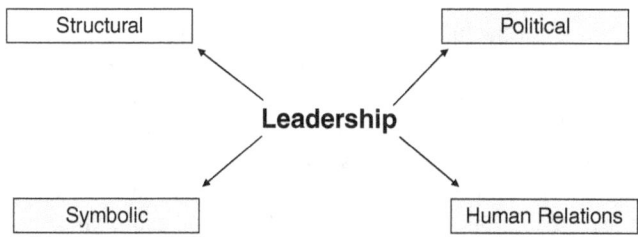

## The Cornerstones: T. J. and Anita Anderson

*People Liked Him*

People liked him not because of his athletic fame,
Tho' he had won awards in athletic games
His was not a brilliant style, His was not a forceful way
But for all he had a smile and an encouraging word to say.
Honest, straightforward, brave and true, Thus he lived from day to day
Doing what he found to do, in a dynamic sort of way.
He kept his friends throughout the years, never rash in word or deed
Never belittled others with snears, but was there in time of need.
He kept school open day and night, with no personal gain
Simply lived and simply died, but the passing of his name
Left a sorrow far and wide. (Anita Anderson, April 1978)

I cannot say enough about Mrs. Anderson. She is a huge part of who I am today.... Who knows where my path may have led had Mrs. Anderson not been a stop along the way, but I do know that the trip would not have meant nearly as much. (J. Diana Lewis, former student)

Just as a cornerstone joins two intersecting walls and is indispensable in the stability of a building, T. J. and Anita Anderson were the metaphorical cornerstones of James Adams, connecting the school to the community it served. Drawing on the metaphor in another way, leadership is also the cornerstone of an organization. Delving into their backgrounds, as well as practice using the HECS model discussed in Chapter 2, we unearth examples of political, structural, and human relations and symbolic leadership qualities and behaviors. To recap, structural and symbolic designations are tethered to individual leader nature and human relations and political aspects of heading a school are ascribed to leader behaviors. Both idiosyncratic leader qualities and leader behaviors contribute to the composite leadership profile advanced in the HECS model.

Though the focal point of the study is between 1943 and 1956, an examination of that time period alone would render an incomplete picture. Mr. Anderson's first appointment between 1925 and 1929 was a pilot period of sorts for the full-service experiment that was implemented upon his return in 1943. By T. J. and Anita's return in 1943, plans to turn James Adams into a full-service location had percolated in both of their minds. As a result, assessment of leadership will include information from both periods the couple worked at James Adams.

It is a bit unconventional, but much of the leadership evaluation considers the contributions of both T. J. and Anita Anderson. These two were a team in every sense of the word both privately and professionally. Interviews confirmed that only rarely did students, community members, or faculty speak of Anita and T. J. Anderson independent of each other. As per one student, "one really cannot look at the Coatesville Community and not see them as a team working tirelessly to make our city and our spirit strong" (J. Diana Lewis). Their daughters confirmed they were an inseparable team, "because Mom and Dad were so dedicated, we were at James Adams School, morning, noon, and night and loved it. We felt loved by the community" (White & Prince, 2007). As such it is important to also consider Anita's structural, political, practical and philosophical contributions to the leadership of the school.

Given both T. J. and Anita's upbringing and academic preparation it is easy to see how the creation of a community school would seem natural and intuitive to them. As it pertains to JACS, the best way to unearth how they came to be such an inextricable and effective pair we begin with the structural aspects of their leadership style.

*Structural leadership*

By definition the structural elements of leadership are preparation, planning, and the experience of the leader in question. Extrapolating those terms further into practical full-service school necessitates an understanding of community challenges; a leader who encourages linking the life experiences of students into the curricula; exemplary academic preparation; and an ability to practically apply educational theory. At their root the best practices of structural leadership ensure that school–community interactions acknowledge and affirm community dignity, mores, and values.

Thomas Jefferson Anderson was born in Lynchburg, VA, in 1889. His father (Jefferson Anderson) was by any standards ambitious. He was one of four Blacks who served on Lynchburg's city council, managed the local Methodist cemetery and his own grocery business with only a common school education. Understanding the importance of a stellar education, he sent T. J. to Philips Exeter Academy, an elite preparatory school in New Hampshire. Jeff (as he was called) probably heard about Exeter and the New England Conservatory of Music (where he sent his daughter Louise) from the White members of the city council (Anderson III, 2000/2001).

After graduating from high school, T. J. worked for a time to support his sister until she graduated from the New England Conservatory of Music and then matriculated at Howard University (Anderson III, 2000/2001). He was an outstanding athlete, having been recognized in "Negro Firsts in Sports" and qualifying for the U.S. Olympic team in the shot put event (Young, 1963). He earned his undergraduate degree from Howard University in 1925 and then assumed the principalship of James Adams School.

As per a student, when he first arrived in Coatesville (in 1925), there was a function where many of the Black men of the community were in attendance. It is important to remember that Coatesville was a steel town and as such was home to men of impressive physical strength. Being new to the community T. J. instinctively knew to be effective he would need to quickly earn the respect of neighborhood fathers. As the story is told, he pointed to the heaviest object in the room (an anvil-like object) and prodded attendees to lift it. After numerous failed attempts, he reportedly walked to the object, not only lifted it but tossed it, and then announced he was the new principal of James Adams (Johnson, 2006). This may seem like an ostentatious display of masculine bravado; however, considered in the context of his general demeanor and low-key personality, one can conclude that the encounter was strategic. The most important element of this story is the recognition that respect for the school leader was in his mind critically important to the success of the school and its students.

Anita Belle Turpeau was the oldest of the ten children of Ila Marshal and David Turpeau. Her father was a prominent Methodist minister with some Ivy League education (Yale Divinity School), a degree in theology from Drew Theological Seminary and was at one point an elected official in the Ohio Legislature. Her mother had matriculated at Gilbert Academy Normal School in New Orleans and was an activist in her own right. Anita's commitment to the marriage of communities and schools was foreshadowed by her parents. Anita and her siblings were guided by many house rules, one of which was "that a group could not be joined unless a public service component was an integral part of the organization" (Current, 2001, p. 35). While her father served as pastor of Calvary United Methodist church in Cincinnati (1923–1933), her mother Ila was on the Board of Directors of the Young Women's Christian Associations (YWCA), was one of the founders of the local Urban League, and was active in the NAACP and Republican Party. During Turpeau's tenure the church became a center for educational, social, and political activity (Current, 2001; Turpeau Sr., Unknown; Kelly, 2008). Both Anita's and T. J.'s sense of responsibility to community and devotion to education were evidently instilled early in their lives.

Anita's decision to attend Howard University was in part prompted by a desire to stay near family (her father was the pastor of Mt. Zion United Methodist Church in Georgetown) and partially because she received a full scholarship. Anita and T. J. (who she called Andy) met when they both matriculated as undergraduates. A short time before the end of her last semester, he told her that ten days after graduation he was "throwing his hat into the ring" as a suitor. He indicated he wanted her to enjoy her time in college but now he wanted her to know of his romantic interests.

Distinguishing herself academically, the future Mrs. Anderson graduated with honors and was the first woman admitted to the university's debating society (Kappa Sigma) and to be the editor of *The Hilltop*, the college newspaper. Upon graduation in 1925 with a degree in social studies education, she took a job as a principal of a laboratory school in Pine Bluff, AR. During her year in Arkansas (postgraduation in 1925–6), she and Andy corresponded, often sharing instructional ideas and administrative strategies. He offered her a job at James Adams School, which she accepted given that it would bring her much closer to her family, who were living in Pittsburgh, PA, and closer to T. J., who by that time was her fiancé (A. T. Anderson, 1982, 1989).

Anita moved into the house for James Adams School teachers that T. J. had established to provide unmarried women a level of privacy. Before the end of the academic year the two were married (April 9, 1927). From that time on, the two functioned as essentially one unit, promoting educational opportunities and meaningfully engaging the community served by James Adams School (A. T. Anderson, 1982).

Overlapping dates suggest that while T. J. was principal he also matriculated at Columbia University in New York, earning a master's degree in Education. For a period that included his first time at James Adams, Columbia, and a few years of his tenure at Howard University, he was also a correspondent for *The Amsterdam News*, a New York City Black newspaper still in print (1926–32) (Chester County Historical Society Archives, January 26, 1957). Given his graduation in 1929 it is very likely that he was a student of John Dewey; at the very least Dewey's influence at that time heavily infiltrated the department. After completing his graduate degree, he was for a short time the principal of Attuck High School in Indianapolis, Indiana.

He welcomed the chance to return to his Howard University alma mater. While a professor at Howard University, he was the business manager for the peer-reviewed *Journal of Negro Education* (1929–43), also still in print. Anita, not one to remain dormant, pursued graduate education, worked, and nurtured their growing family. By the time they left Washington, she had completed the requirements for two master's degrees, had been the Director of the Wesley Foundation (a Methodist educational loan fund), and was the mother of three children.

T. J. remained a scholar throughout his career, writing for periodicals and reading voraciously, and even into retirement there is ample evidence of how well he was respected in educational circles (an observation substantiated in part by his inclusion in a volume of "Who's Who in American Education"). In 1948, he was a Carnegie Foundation scholar and spent a summer at the University of Nottingham, England. After his retirement, he was appointed by the U.S. State Department to serve 2 years as an educational envoy to Iran to aid in the establishment of their educational system (1957–9) (Chester County Historical Society Archives, January 26, 1957, October 19, 1948).

As for Anita by the time she completed her career as an educator, she had earned two master's degrees (in Education and Religion) and a law degree (La Salle University). Anita herself was a prolific writer, reader, composer, accomplished musician, and scholar. Her works as a sample have been included in National Anthology of Poetry and the journal *Avatar* (A. T. Anderson, 1982, Unknown-a; Chester County Historical Society Archives, April 1949).

Though professional pursuits lured the Andersons away from Coatesville, their hearts remained in the little town. Anita recounted that while at Howard, T. J. constantly talked about Coatesville (A. T. Anderson, 1982). The call to return to Coatesville came in 1943 after a fire destroyed the entire library collection and the school reopened in 1942. As per Mrs. Priscilla Stokes, a former student and teacher at the school, people were thrilled to have the Andersons back (Cuff-Stokes, 2007).

When this impressive pair returned to Coatesville, they were well equipped to implement their vision of how a school could function as a community hub. To recap, their undergraduate education—though garnered from a historically Black university (HBCU), Howard—was deemed highly respectable by Whites because of its reputation as only one of two HBCUs at the time capable of meeting the standards for accreditation established by the Carnegie Foundation. This assessment was promulgated by Thomas Jesse Jones, who at the time was the director of research for the Phelps Stokes Fund (J. D. Anderson, 1988). By 1943, they had between them three master's degrees (one from Columbia University), experience in principalship, research and community-organizing expertise, and intimate knowledge of the idiosyncratic challenges and opportunities of the city. Ultimately, they came back to Coatesville with a plan for a viable community school, or in their words, "a school by day and community center by night."

Partnerships had to be the heart of the community–school relationship. Collaborations with the school were deemed as necessarily mutually beneficial. Stated another way, affiliation with JACS should not only benefit the school population through the provision of resources, but also be viewed as a good investment by the citizens of Coatesville. It is also apparent that the Andersons were not presumptuous about the priorities of families. For example, community surveys were periodically circulated asking for input which kept the Andersons and the rest of the teaching staff of shifts in community priorities (Chester County Historical Society Archives, November 13, 1946).

A multitiered approach best sums up the comprehensive Anderson strategy. Mr. Anderson's membership on the Interracial Committee afforded him access to city council members that his role of principal would not necessarily afford. He used the exposure to identify opportunities for students and faculty to participate in the political, social, and cultural life of the city. In addition, as a member of the Interracial Committee, he participated in a study of the ways in which racism detracted from the life of the town as a whole and the African American community in particular. The report was issued with findings that influenced the willingness of majority-owned businesses to invest in James Adams School for years to come.

The committee also created platforms to highlight the contributions made to Coatesville by Adams and articulated the ways in which the city's political actors could be more responsive to the needs of the African American community. The committee's regular publication, *Concern*, was used as a dais to publish Adams accomplishments and announce upcoming events held at the school or in other venues of cultural significance, such as a Leontyne Price concert. Businesses and organizations were regularly thanked publicly and privately for engaging in more racially inclusive practices (Chester County Historical Society Archives, December 31, 1945, February 23, 1944).

Mr. Anderson's principalship the second time around provided almost immediate demonstrations of the direction the school was about to take. Whether a function of his reputation or active recruitment, he assembled the most highly qualified faculty in the district among White or Black schools. At one point, every teacher in the lower school (K–5) had a master's degree (Cuff-Stokes, 2007). Mrs. Anderson concomitantly engaged her circle of civic and club memberships as an opportunity to both serve the community and identify support for the school. Her creativity expressed through her prolific writing and notable musical skills were also a bridge between the school and the community.

Student interviews in combination with a survey of Adams programs attest to the ways activities and curricula were complementary. Proper grammar use was a skill perpetually reinforced by Mrs. Anderson in all activities. As per one student, "I remember being on the radio (WCOJ in Coatesville). There was also a time we were on television—WHYY, Wilmington, Delaware. Somehow, I got my line mixed up and had to rearrange the working of the sentence. Thankfully I got it right. On the way home Mrs. Anderson said, 'I was afraid you were going to split a verb.'" A more detailed account of the types of programming that linked curricula to activities for students is found in Chapter 5, including some events that were annual (such as operettas, the winter carnival, oratorical contests, and graduation panels and school newspaper submissions). In the case of others, it is more difficult to determine if they evolved over time or were in place for singular academic years. However, what is clear from interviews an newspaper articles referencing the selected school years is an organic and eclectic mix of courses and activities that were tailored to the changeable priorities of the community.

Reflecting back on the Andersons, it is clear that their academic and life experiences informed their notions of what JACS could become. Borrowing from the proverbial "leaving no stone unturned," they used every venue available to them to enhance the life of the school. In the vision statements they published and ethos they espoused, the school was inextricably part of the African American community and the Coatesville fabric. They believed that citizens of any hue would benefit when commingling in social, cultural, and educational settings. Isolation threatened not only Blacks but Whites from developing to their full potential. As such, more than being a school, JACS modeled how integrated engagement might function. Given the priority placed on partnerships as outgrowths of relationships it seems natural to turn our attention to the next dimension of leadership espoused in the HECS model—human relations.

### *Human relations leadership*

Human relations in the context of leadership speaks to the types and qualities of relationships the leader builds. Authentic engagement with individuals and organizations cannot be grounded in self-serving agenda. Partnerships born of a selfish ilk tend not to be sustainable nor are they the most effective. Like friendships, partnerships have a type of mutual benefit associated with them, so a delicate balance must be struck as supportive networks are created. Restated, as leaders identify strategic partners they should be thinking not only in terms of identification and engagement but also the benefits that will accrue to students, the community at large, and the partner specifically. Some of the characteristics of effective relationship building are that the leader is perceived as welcoming; s/he spots opportunities in conventional and unconventional collaborations; and s/he possesses communications skills that support interpersonal relationship building. T. J. and Anita were clearly proficient in this arena.

Defying notions of elitism and professional turf, the Andersons actively created programming that met the needs of the community by first soliciting suggestions from staff and individuals external to the school. After-school classes given the attentiveness to input were more apt to garner support from partners because their value had already been established. Adult evening offerings included social, civic, and educational courses. Samples spanned securing General Education Degrees (GED), driver's licenses, completing taxes, first aid classes, and applying for citizenship (Chester County Historical Society Archives, October 18, 1950). Collaborators valuing these offerings supported JACS with volunteer hours, finances, and in-kind contributions. In turn, individuals who benefited from classes possessed more power over their own destiny, were endeared to the school, and were willing to volunteer their time and resources.

It could be said that the commitment to reciprocity and the empowerment of students and their families was the philosophical underpinning of partnership development. Also pivotal was identifying whether discrete investments or sustained engagement were the best entry points for initial collaborations. As a result, the list of individuals and groups T. J. and Anita sought to engage was long and strategic. It seems the Andersons thought carefully about the reciprocity that might appeal to partners. Teachers, custodial and office staff, parents, social clubs, civic organizations, and business were all potential partners with unique sets of collaboration motivators. Attention to what allies valued in a spirit of Can I Help You (CIHU), which will be discussed later, seemed to have been additive to the Andersons' efforts.

It is important to remember that the work of JACS was accomplished primarily as a function of volunteerism and donations. Teachers and staff were viewed as partners in the work as well. In the absence of these relationships, the work would never have been accomplished. Major assistance came from the Young Men's Christian Association (YMCA), whose support took the form of funding, workers and supplies. Shortfalls were covered initially by the Andersons and other teachers and staff.

Surveys, interviews, and newspaper articles paint a picture of perpetual exchanges between and among collaborative partners. Parents who were gifted cooks supported the school and community functions. Seamstresses made costumes, vocational classes were led by artisans, and students' products were used either by the school or the community. For example, the furniture produced by the woodshop classes were so well made, that it rivaled the quality one could buy in local furniture stores.

The Andersons' genuine love for the people of Coatesville drew individuals and resources to them. Part of their leadership mantra was to lead by example. There was nothing they asked of any partner they were not already doing. If the school or their home was the most appropriate site for a function, it was made available. Their expertise was of value beyond the school community and as such it was not uncommon for Mr. Anderson to be called upon to serve on boards, as a business advisor, or in other capacities. Anita, like her husband, led numerous civic organizations and in addition was an accomplished pianist and was frequently requested to play at church, community, and business functions.

Though JACS was a segregated school, it was not uncommon for persons of all hues and professional stations to aid in raising money and strategizing how to best meet the needs of students. As per Mrs. Cuff-Stokes (2007), "Whites would come to Adams because the Andersons were so well thought of in the community." One such person was Rev. James Brown, a White Presbyterian minister who was quite influential and one of T. J.'s dearest friends. Anita's friendships also crossed color lines as she often referred to Margaret Thompford, a White business owner who was also a Quaker, as one of her best friends (A. T. Anderson, 1982). Race, religion, or community standing had no relevance to the Andersons. What mattered was the character of the individual and their willingness to enter into relationship. There were those who were partners to the school and there were others who were partner-friends of the Andersons (by partner-friends I refer to individuals who not only contributed to the life of the school, but were guests in each other's homes). In either case, T. J. and Anita were adept at matching interests with opportunities at JACS.

Relationships also were the foundation for other tangible community support. For example, one of Mr. Anderson's friends (who also graduated from Howard University) was Dr. Whittier Clement Atkinson. Atkinson Hospital was a major recipient of the melded reciprocity and CIHU commitment of the Andersons. After witnessing the treatment of his patients (in 1925), Dr. Atkinson opened one of seven hospitals in Pennsylvania devoted to Negroes in 1936. The hospital was blocks from JACS in the heart of the East End African American residential enclave (Chester County Historical Society Archives, April 15, 2007). Reciprocity relative to Dr. Atkinson's relationship with JACS was rendered through T. J. serving on the hospital board as director of public relations, securing patients for his practice and hospital, and the many school-initiated drives that provided food and finances for the facility. Dr. Atkinson and T. J. arranged to have a medical and dental clinic in the school where inoculations and other services were administered on-site. Provision of this battery of services is necessary for a school to be considered full-service in the contemporary lexicon (Dryfoos, 1994).

School publications such as *The Adamite* and *The Adams Gazette* contain significant evidence of local business investment in JACS. The longer the Andersons lived and worked, the more support of all kinds was made available to students. Several interviewees commented on how it was impossible to go anywhere in Coatesville with either T. J. or Anita and not be stopped (Cuff, 2007; Garnett, 2008). Those relationships served students and faculty in terms of not only fiscal support but also positioning the school for purposes of political support. The ways the politics of leadership and resource distribution can be welded to support an agenda will be discussed next.

*Political leadership*

In relation to leadership, political attributes are demonstrated when a principal exerts his/her influence to ensure resources are distributed in a way that supports the mission of the school. Politics of leadership also refer to the ways leaders share, delegate, and exert power. Some are more authoritative while others are more interested in creative networks of control that allow for flexible roles of staff and volunteers. Highly effective political leadership avails itself of both ends of the spectrum depending upon the situation. The latter categorization was Mr. Anderson's preferred approach. While there was no doubt that he was the principal and final decision maker, he walked the delicate line of autocracy and democracy within his school. He was ever present in the hallways, before and after school. Yet, he felt no need to be "up front" during events. Students were clear about his expectations as they remembered, "He was fair but tough." Another student recounted, "He would not let you stray from [prioritizing] education."

Broader conceptions of politics incorporate the decisions determining the allocation of resources and how they are used. One way to check for political clout as proposed by Jago (1982) is to investigate how power is accessed and applied. Power affords individuals and organizations entry to beneficial resources, networks, and influence. The Andersons by all accounts enjoyed power in both their own and the White sectors of Coatesville life (Cuff, 2007; Garnett, 2008). Once attained, power can be operationalized in a way that is self-serving. The Andersons, however, seemed to perpetually seek opportunities to use their influence to the benefit of others.

One of the first questions is how in a segregated era did a Black couple commandeer such influence. Clues lie in the Andersons' professional and personal affiliations; their structural leadership attributes; cadre of influential allies; and the positive contributions of the school. However, that alone does not explain the enduring respect he and his wife command decades after their deaths. Among Coatesville residents influence was built interaction by interaction. Mr. Garnett fondly recalls the many times the Andersons' kindness crossed race, ethnicity, and social class, making the couple beloved by many (Garnett, 2008). One of Anita's sisters who visited often noted that T. J. and Anita were "Coatesville royalty" (Kelly, 2008).

Building blocks of influence and respect were earned not only by virtue of successful educational interventions, but also every time the Andersons quietly met the needs of families. It was commonplace for T. J. to anonymously drop presents and food off to families. Though it was his intention to slip in and out without being detected, given his stature (he stood over six feet tall with a football-like physique) that proved quite the challenge. Aid and encouragement were provided in many forms. A student who stayed with the couple for a period while her parents handled family business in North Carolina and another whose longing for a pair of little red shoes that mysteriously appeared under her Christmas tree both feel a sense of gratitude a half century later.

The Andersons' home was a way station for persons needing shelter, counsel, comfort, or food and clothing. Recipients were not necessarily connected to the school; however the reputation for genuine care and concern for persons accrued to the school in tangible and intangible ways. When the Andersons sought volunteers for after-school programs, community members welcomed the opportunity to give as much as a way to repay T. J. and Anita as any other motivation.

Power and influence in the Anderson ethos seem to serve to as a springboard for efficacy. For example when the school population exceeded the building capacity T. J. petitioned the city school board to build an annex and initially he was denied. Undaunted, T. J. launched a public relations campaign to support his effort to which the school board soon conceded (T. J. Anderson, 1949).

Part of T. J.'s influence seemed also to stem from his affiliations. As a member of the gubernatorial authorized Interracial Committee's reporting, public forums, and publications, he was able to garner local business community support for school functions and scholarships from those desirous of positive public relations and those of like mind. Looking back, Anita's reputation as a scholar-educator and community leader in her own right was acknowledged in numerous ways. She was one of 100 African American women invited in 1938 by Mary McLeod Bethune to the White House Conference Group hosted by Mrs. Eleanor Roosevelt to discuss the state of education for Blacks (Willis, 1992). She was the first African American teacher hired at Scott High School after JACS was closed and in 1974 was inducted into the Coatesville Hall of Fame (Chester County Historical Society Archives, February 11, 1974).

Another example of the Andersons' influence can be found in the many high-profile individuals who would come to JACS. Famed activists, concert singers, authors, actors, and athletes such as Paul Robeson came to the school. It is not clear when he and T. J. met; however, over the years they became close friends and as a testament to T. J.'s steadfast integrity, even when Robeson was censored and forbidden to travel abroad, T. J. took his family to one of his concerts in Philadelphia. Jean (T. J. and Anita's youngest daughter) recalls not only the concert, but the intense conversation the two men had backstage after the concert (White & Prince, 2007). Other persons of note included Roland Hayes, the first African American to win acclaim in the United States and Europe as a concert artist, and Leontyne Price, the celebrated concert soprano. Sports notables such as Wally Triplett were also known to come to the school. Triplett was the first African American to be drafted by and play for the National Football League (NFL). It was not uncommon for the college presidents of Cheyney State Teachers College (a local HBCU) such as Leslie Pinckney Hill to address JACS students even though they were years from matriculating at the postsecondary institutions. In these cases, the Andersons tapped their networks to provide access to role models for students.

There were also strategic mentors\allies who by virtue of being political actors did much to bolster the Anderson's ability to bring to fruition the vision of a fully functional community school. Political actors are individuals who by virtue of their position or contacts weld influence in matters of politics, such as in this case the distribution of school board assets to JACS.

One such person was H. R. Vanderslice, who was the superintendent of Coatesville schools during T. J.'s first tenure at JACS. Vanderslice was a lifelong ally to JACS, who despite the constraints imposed by segregation remained committed to advancing equality educational experiences for students of all color in Coatesville. He disputed the rationale for segregation in more than one report to the school board but apparently was forced to acquiesce to separate schools for Blacks and Whites. In his 1923 report, Vanderslice attributed the types of experiences students had after they left segregated school in sixth grade as one culprit in the underperformance of Black children that was not the fault of the children or teachers. It is not clear what additional factors he considered, for example, whether or not, knowing that garnering support for equitable treatment of Negro children in White schools would not yield aid; however, what was clear was his goal of ensuring an improved graduation outcome.

Under his leadership JACS received more fiscal resources, facilities improvement, more faculty with high educational credentials, and he increased the scope of the school to include up to ninth grade (Chester County Historical Society Archives, December 1950a). He worked with

Mr. Anderson in order to ensure to the facilities, staff, and fiscal support were aligned as closely to the White schools as possible. Even after he left Coatesville he remained associated with JACS, returning from time to time as a keynote speaker at events; in 1955 he was awarded one of JACS's highest honors, the Americanism Cup for his commitment to the mission of the school (Chester County Historical Society Archives, May 28, 1955).

In 1950, Superintendent Cecil Rice made recommendations for a "sound" educational system in Coatesville that did not support parallel schooling. It appears such a report was prompted by overcrowding at JACS. He advanced the options of integrating the seventh through ninth grades of Adams and Gordon (the White junior high); building a new high school to house Coatesville and suburban students (his preferred plan); or building an annex to James Adams to accommodate growth and leaving the seventh through ninth grades located where they were (Chester County Historical Society Archives, December 1950b). Impetus for his preferred strategy was grounded in his closing remarks of the report, which read,

> Adding more rooms [to JACS] would mean that the Board intends to perpetuate the policy of race segregation in our junior high schools. Such a policy brings great problems. Segregation is expensive because we must give the Adams Junior High School children the same educational opportunities that we do the Gordon Children.... It is expensive in terms of his [students'] adjustment too, because Adams children get to the tenth graders—15 or 16 year of age—before they ever get opportunities to adjust their lives to living and working closely with [W]hite children. Also the reverse is true, and the [W]hite child often has just as much difficulty in making the adjustment. The races should be integrated at a much earlier age in our public schools for the social good of all concerned. (Chester County Historical Society Archives, December 1950b, pp. 3, 6)

Important to remember here is the value of a powerful political ally advocating on behalf of your institution. Even though the city politicians were not ready for full-scale desegregation, Rice's stance provided documentation of support and eventually, though not the optimal solution, additional resources and support for JACS programming. The Coatesville School Board still did not fund JACS equitably when compared to the local White schools; however, the combination of T. J. Anderson's high profile, access to media, and skill as an advocate, and well-placed political advocates laid a pathway for greater support than would have otherwise been available. Remaining gaps in funding were then met by the generosity and hard work of the partnership network.

The purpose of political influence in the Andersons' world was to empowering people to meet their own potential. As the priorities and challenges of families changed, so did the programming, which brings us to another lynchpin in their successful creation of a community school—their symbolic leadership.

*Symbolic leadership*

The final category of leadership to be explored relative to T. J. and Anita is symbolic, or an ability through commitment, tenacity, and innovation to communicate a vision and promote the adoption of the said vision by those who are central to implementation. Literature speaks of the ability of a leader to accomplish the goal of facilitating the adoption of a vision as requiring personal commitment, charisma, and an ability to motivate others as a servant leader (Popper, 2000). About this concept one student noted, "[l]ately, one of the biggest catch phrases in business is 'servant leadership'. Businessmen and women around the country are teaching that you can't lead effectively if you don't also serve. I laugh at this 'new' concept. These people obviously didn't know Mr. Anderson. He was way ahead of his time."

Charisma and personal commitment tend not to be acquired skills. T. J. and Anita each possessed a magnetism and charm that drew people to them. Charismatic leaders are depicted as individuals who arouse devotion and this was clearly the case with the Andersons. Another characteristic of symbolic leadership is personal commitment. It is difficult to provoke dedication in others when as a leader you do not demonstrate dedication yourself. Commingling of the personal and professional lives of the Andersons demonstrated daily to members of the Black and White communities that they were vested in the success of not only the school but also individuals from all walks of life.

There was virtually no demarcation distinguishing where their personal lives ended and professional lives began. Their home was as much a community hub as the school. Thus part of the trust that was built between the Andersons and the community lay in the demonstrative consistency of their commitment. This synthesis of private and public personas was also demonstrated in the ways the faith community was embraced in the school and the ways the Anderson's moral code directed their work. Coatesville enjoyed great religious diversity at the time. Presbyterians, Methodists, Baptists, Catholics, Seventh Day Adventists, and Quakers all had churches within or just outside the city limits. T. J. and Anita welcomed persons of all faiths into their home and to the school, which yielded rich multicultural experiences for faculty and students. Just as one cannot separate the professional and private lives of the Andersons, likewise it is futile to attempt to compartmentalize their faith. They were not enigmas of the time because religious-like zeal of Negro teachers during this period has been compared to a "calling" (Walker, 1996). In keeping with this rationale, the devotion to the integrity of the work logically embodies the devotion of the educators, as reflected in one of Anita's poems entitled "The Teacher's Prayer":

> As teachers, we thy guidance should seek,
> Give us instruction on how to speak.
> Acquaint us with the golden rules,
> Be thou head master of our school.
> We offer you our youth,
> Lead them to eternal truth.

The power of their example proved a significant motivator, more so than anything they might have said. Anita was a prolific poet and would often publish inspirational sayings and poems in mainstream press, *The Adams Gazette*, and booklets that were passed on to students. One such booklet entitled "Just an Earful" was given to me by an alumnus, who upon hearing this book was being written wrote, "I have probably had this [booklet] for fifty years. It belongs in your family." Before turning to accounts of included expressions, one can well imagine the significance a little 14-page booklet must have had to be retained for 50 years. Sample thoughts in the book are "Go forward; let yesterday go" and "Move into the now with determination and foresight"; "Attitude must be watched in order to move to an altitude of happiness"; and poignantly, "We can be surface people or we can get acquainted with ourselves and operate on high dimensions" (A. T. Anderson, Unknown-b). Speaking of motivation, any evaluation of the Andersons' symbolic leadership would be incomplete if it did not include the ways CIHU infiltrated the Adams community.

What is CIHU exactly and how did it become part of the school philosophy? CIHU stands for Can I Help You. Embedded in the acronym is the notion that helping the community is the mission of not just the school but citizens in general (Figure 4.1). As was noted earlier, the Andersons were persons of faith—Anita a lifelong Methodist and T. J. a Seventh Day Adventist. On one of her frequent summer religious retreats, Anita attended a Camp Farthest Out (CFO) retreat with Margaret Thompford. CFO retreats are nondenominational gatherings of those interested in deepening their personal spirituality. In a meeting dating back to the mid-1940s, Anita was exposed to Frank Laubach's CIHU characterization of service. Dr. Laubach was reknowned for his work as a literacy missionary who advanced the Each One Teach One program around the world. As part of that initiative, he proposed Christian service should embody a daily CIHU practice (Chester County Historical Society Archives, June 1954).

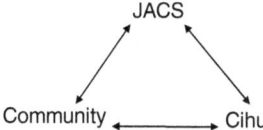

**Figure 4.1** CIHU and the Community.

The earliest uses of the acronym in the school can be found in 1946, though the ethos it embodies had been evident long before its formal adoption. Anita's ability to bring CIHU from an ideal to practice caught the attention of a reporter (Amos John Travers) from the *Christian Herald*, a magazine with national circulation. In his article he described Coatesville and JACS as "a busy industrial city whose public schools are a melting pot of many races which is regarded as a splendid laboratory for testing the theory [CIHU]" (Chester County Historical Society Archives, June 1954, p. 13).

Dr. Laubach heard about CIHU taking on a life of its own in Coatesville and requested an article be written about the school, which was first published in 1946. Additional articles were written in miscellaneous publications after that date. Dr. Laubach invited representatives of the Adams faculty to Koinonia College near Baltimore to aid in training others in the ways CIHU can be translated from a philosophy to action (Chester County Historical Society Archives, October 1953). In addition, students wrote letters to Dr. Laubach explaining what CIHU meant to them, some of which were published in syndicated articles (Chester County Historical Society Archives, June 1954).

Years later when she thought of CIHU in action Anita began with examples from JACS. She referred in one interview to a young man who upon arriving at JACS had a reputation for being "incorrigible." After transferring to Adams as a last attempt to retain him in school, he was assigned leadership responsibilities and became over time a trusted member of the school patrol. This she thought spoke to the power of CIHU. Mrs. Anderson recounted how rewarding it was that "people say to me on the street, 'I've never forgotten CIHU'" (Hungerford, February 11, 1974).

Interviews with JACS students confirm that CIHU was more than a school motto; it permeated the life's practices of alumni. "To me, CIHU is a way of life," one alumnus wrote. Another said of CIHU's influence, "it encouraged me not only to achieve in my own right, but it encouraged one to help others achieve." As further testament of its' (CIHU's) symbolic power, alumni have perpetuated the tradition by investing in their communities as adults. As an association JACS alumni continue the CIHU tradition by awarding multiple scholarships to African American students graduating from Coatesville High School. For example, depending on donations, as many as five $500 scholarships have been awarded at the annual reunion. This ongoing commitment is even more impressive given the age of alumni (now ranging in the seventies and eighties), most of whom are on a fixed income but manage each year to contribute enough to help several students (Johnson, 2006).

As for the Andersons' willingness to take risks (a hallmark of effective symbolic leadership), it should not be overlooked how intertwined their approach to living and working in Coatesville was. They defied convention and assumed that many Whites secretly supported integration, but did not have safe environments to "practice" interacting with Blacks. This absence of exposure was not considered a personal flaw, but rather an opportunity for JACS to provide a safe haven for interracial mixing and advocacy.

Given the historical context, there were distinct gender-based expectations. However, early in their tenure, the Andersons determined whoever could get the job done most efficiently should be in charge. As a result, women assumed leadership roles in the school by day and also in the evening programming. To further press gender expectation, T. J. (in what by a newspaper account was deemed quite extraordinary) allowed girls to participate in classes that historically had solely been offered to boys, such as woodshop. While evening programming for adults was available to both genders, girls tended not to have as much access to evening activities as boys, that is until T. J. offered student sessions for boys two days a week, girls two days a week, and on Fridays, all activities were coeducational (Chester County Historical Society Archives, October 10, 1945).

An ability to articulate a clear vision in collaboration with partners is one of the final benchmarks of effective symbolic leadership. Evidence can be found in many venues beginning with the reports prepared for the school board where Mr. Anderson's sentiments are clear. In one report he states, "the ultimate objective of modern education is good citizenship. It is designed to produce happily adjusted personalities. . . . It demands thinkers rather than learners" (T. J. Anderson, November, 1953, p. 4).

In another publication Mr. Anderson published six points as the guiding principles of JACS, which embodied his vision for students and the extended community:

1. Education (Acquisition on the part of pupils of the rudiments of liberal education through contacts with and participation in those things which represent the accumulated wisdom of the human race together with the widest possible opportunity for exploring the fields of knowledge). The education program centers around three points of emphasis

a. Guidance which conceives of the child as a total personality.
b. Classroom activities devoted to the formal acquisition of skills, knowledge, habits, appreciation, etc. as prescribed by the Pennsylvania State Department of Education.
c. School life activities are laboratories for discovering and expressing pupil interest.

2. Health—A sound mind in a sound body. Developed through classes in physical fitness, health and recreation and intramural programs.
3. Guidance—Including vocational and social adjustment, self realization with reduction of delinquency as a corollary.
4. Community Organization—People integrated through common experiences conscious of local unity; developing basic service institutions; and able to act in a corporate capacity.
5. Recreation—Worthwhile activities participated in for their own sake providing relief from tension, satisfying the desirable drives of original nature, and promoting maximum personality development.
6. Race Relations—Recognition of the positive contribution which all Americans have made and are making to human welfare, elimination of racial stereotypes; interpreting to the community the legitimate aims of Negroes in American society, and promoting mutual goodwill and self respect. (Chester County Historical Society Archives, September 1943)

Practical demonstrations of the published points informed JACS in its totality. Aligning JACS's vision with practice was demonstrated in the commitment to hire and empower highly qualified teachers in the school. Health and recreation were supported through school-based clinics, physical education, and extracurricular sports programming and May Day exercises. Recreation also included clubs and after-school fine arts programming (drama, orchestra, arts and crafts). Personal, educational, and career guidance was provided through one on one interactions. Regularly persons representative of diverse backgrounds and thought were part of programs and assemblies. Students were also taken out of their neighborhood enclaves as much as possible to increase their exposure, thus nurturing the skills they would need to function well in an integrated society.

As an aside, the organization of the school newspaper itself was unique in that articles included intensely scholarly contributions and inspirational and thought-provoking prose, alongside typical school newspaper articles recounting sporting and social events tied to the school. The same edition hosting scholarly contributions entitled "The Negro in Coatesville", "Early

History of Coatesville," and "One Objective of Education", contained prose interspersed throughout, such as "A smile costs nothing, but gives much" or "Obstinacy is the ego's peacock plumes." The publication was complete with volume and number designations akin to peer-reviewed journals of today (*Gazette*, 1950).

The philosophy undergirding the Andersons' approach to schooling, in the words of T. J., was "[f]irst, to raise the cultural norms of the people it (the school) served through an all encompassing program geared to meet total community needs not adequately supplied by other societal agencies; secondly to move from parallelism towards complete integration; and, thirdly to make democracy work in a practical level locally framed in an understanding of the Declaration of Independence and the U.S. Constitution" (Anderson, 1950, p. 5). Armed with the clarity of their vision, T. J. and Anita set forth to impress upon students, teachers, and family members the centrality of JACS to their hopes and aspirations.

## Comparing Leadership Then and Now

Leadership is a topic of great interest across disciplines, thus providing a wealth of research upon which to draw a composite sketch of effectiveness. Qualities deemed valuable include innate, observable, and

Table 4.1 Community School Leadership Comparison

| Leadership Category | Best Practices | T. J. & Anita Anderson |
| --- | --- | --- |
| Structural | Understands community challenges | Graduate preparation in education (both)<br>Experience as a principal (both)<br>College professor (T. J.)<br>Periodical business manager (T. J.) |
| Preparation Planning Experience | Encourages linking life experiences to curricula<br>Exemplary academic preparation<br>Ability to employ strategies appropriate to the community context | Author (both)<br>Teacher (both)<br>Initial appointment exposed couple to community norms, practices, challenges, and opportunities (both)<br>Determining the type of community school Coatesville needed, i.e., medical services, programming in consultation with community, scheduling (both)<br>Purposeful alignment of curricula to after-school activities (both) |

| | | |
|---|---|---|
| *Human relations* Collaborative partner identification | Perceived as open/welcoming Adept at working with interpersonal and inter-organizational dynamics Strategic partnership/ally development | Partners are broadly defined Community at large Teachers Business community Community of faith Educational establishment Diverse professional and personal relationships to include Whites, immigrants, etc. Creative development of mutually beneficial partnerships Strategic alliances to accomplish specific tasks |
| *Political* Influence Power Resource allocation | Information sharing Resource mingling Defying professional turf Sharing power Knowing the community power brokers | High-profile visitors and performers Fluid programming Broad-based fiscal investments for school and students (school annex, library) School publications |
| *Symbolic* Vision Commitment Tenacity Innovation | Charisma Personal commitment Articulating a clear vision An ability to motivate others Risk taking | Personal and professional lives commingled CIHU Engaging sponsorships and patrons as partners Student scholarships Use of periodicals and publications to advance the school mission |

*Source*: Blank, 2004; Dryfoos, Quinn, & Barkin, 2005; Kouzes & Posner, 1992; Popper, 2000; Rice & Harris, 2003; Riehl, 2000; Sanders, 2001.

learned skills. Table 4.1 shows the qualities of the highly effective leaders (structural and symbolic) and leader behaviors (political and human resources); literature-based characteristics tied to best practices; and, in the final column, specific examples as related to the Andersons.

How do the backgrounds and leadership styles of the Andersons compare with literature-based benchmarks for structural leadership excellence? Without recounting the entire text, Table 4.1 summarizes how in retrospect there is evidence that the Andersons possessed leadership qualities and employed behaviors that support a highly effective community school. Leaders need to inspire confidence and a desire to align with the articulated vision for the organization (structural and symbolic). This is accomplished in part as a function of preparation, modeling the way to demonstrate commitment to the ideal and enlisting the support of others (structural, human relations, and symbolic). Principals as leaders advancing community schools as a strategy to improve student outcomes need to be comfortable with a certain degree of risk and employ flexible leadership styles that distribute decision-making power (political and symbolic). Particularly in the case of full-service schools, identifying partnerships and opportunities that are mutually beneficial is a critical skill (human relations). In addition for a community school to be sustained over time, it is important to acknowledge incremental gains, as well as regularly celebrating contributions to the life of the school (Camburn, Rowan, & Taylor, 2003; Jago, 1982; Kouzes & Posner, 1995).

What becomes most evident in this examination is the Andersons' penchant to lead by example. Kouzes and Posner (1995) refer to this as "modeling the way." Just as children do not function in isolated and unrelated quadrants, neither do leaders. There seemed to be nothing the Andersons asked others to do for JACS that they were not doing themselves. Demonstrating their willingness to spend their time and resources to support the school set an example for others and convinced naysayers of their authenticity.

Masterful identification of the need for reciprocity in partnerships explains how Mr. Anderson commandeered attention and resources to the school. Using his (and his wife's) positions and civic organizational membership, support of political actor advocates, and community relationships, he demonstrated strong leadership capacity (political and human relations). By weaving said relationships into the fabric of the school they were able to expose students to renowned artists, scholars, and use their position to gain more support than might have otherwise been allocated given the segregated school funding structure.

T. J.'s and Anita's charisma independently was only multiplied when they worked together. That charisma served as an incubator for innovation and engagement. If a school is to be the hub of the community, an "as-if" haven, its leaders need to be trusted. Building blocks of that trust are preparation, experience, and the development of plans sensitive to and informed by the community and the other dimensions of leadership clearly demonstrated by the Andersons.

### References

Anderson, A. T. (1978) *People Liked Him*. Personal journal.
Anderson, A. T. (1982). *Interview*. Coatesville, PA: David Turpeau.
Anderson, A. T. (1989). *Interview*. Christiana, PA: Angella T. Hayes.
Anderson, A. T. (Unknown-a). The house on Cedar Street. *Avatar*, 3(23).

Anderson, A. T. (Unknown-b). "Just an earful." Coatesville, PA: Self-published.
Anderson, J. D. (1988). *The education of Blacks in the south 1860–1935*. Chapel Hill: The University of North Carolina Press.
Anderson, T. J., III. (2000/2001). A name reclaimed: Four generations of Thomas Jefferson Anderson. *Lynch's Ferry*, 26–28. Lynchburg.
Anderson, T. J. (1949). *Letter to Superintendent Cecil L. Rice*. Coatesville, PA.
Anderson, T. J. (November, 1953). *Coatesville City School district minutes*. Coatesville: Chester County Historical Society Archives, Coatesville City School District.
Blank, M. J. (2004). How community schools make a difference. *Schools as Learning Communities, 61*(8), 62–65.
Camburn, E., Rowan, B., & Taylor, J. E. (2003). Distributed leadership in schools: The case of elementary schools adopting comprehensive school reform models. *Educational Evaluation and Policy Analysis, 25*(4), 347–373.
Chester County Historical Society Archives. (September 1943). *Our schools in action: James A. Adams school* (Vol. submitted by T. J. Anderson). Coatesville, PA: Coatesville City School District.
Chester County Historical Society Archives. (February 23, 1944). *Piano recital Philippa Duke Schuyler*. Coatesville, PA: Coatesville Record.
Chester County Historical Society Archives. (December 31, 1945). *A song recital: Roland Hayes*. Coatesville, PA: Coatesville Record.
Chester County Historical Society Archives. (October 10, 1945). *James Adams Community School*. Coatesville, PA: Coatesville Record.
Chester County Historical Society Archives. (November 13, 1946). *James Adams Community School*. Coatesville, PA: Coatesville Record.
Chester County Historical Society Archives. (October 19, 1948). *James Adams Community School*. Coatesville, PA: Coatesville Record.
Chester County Historical Society Archives. (April 1949). *Current achievements of Adams faculty*. Coatesville, PA: The Adamite.
Chester County Historical Society Archives. (December 1950a). *The 1923 report by Superintendent H.R. Vanderslice*. Coatesville, PA: Adams Gazette.
Chester County Historical Society Archives. (December 1950b). *Summary and conclusion of the 1950 report by Superintendent Cecil L. Rice*. Coatesville, PA: Adams Gazette.
Chester County Historical Society Archives. (October 18, 1950). *James Adams Community School*. Coatesville, PA: Coatesville Record.
Chester County Historical Society Archives. (October 1953). *Our schools in action: James A. Adams school* (Vol. submitted by T. J. Anderson). Coatesville, PA: Coatesville City School District.
Chester County Historical Society Archives. (June 1954). *Adams home of CIHU*. Coatesville, PA: Adams Gazette.
Chester County Historical Society Archives. (May 28, 1955). *Dr. Harvey E. Vanderslice*. Coatesville, PA: Coatesville Record.
Chester County Historical Society Archives. (January 26, 1957). *T.J. Anderson appointed to post in Iran*. Coatesville, PA: Coatesville Record.

Chester County Historical Society Archives. (February 11, 1974). *Anita Anderson inducted into Coatesville Hall of Fame.* Coatesville, PA: Coatesville Record.
Chester County Historical Society Archives. (April 15, 2007). *Coatesville hospital's founder is honored by historic marker.* Coatesville, PA: Philadelphia Inquirer.
Coatesville City Council. (2005). *Minutes: Monday October 24.* Coatesville, PA: Coatesville Ledger.
Cuff-Stokes, P. (2007). *Interview.* Coatesville, PA: J. W. Richardson.
Current, A. (2001). *Breaking Barriers: An African American Family & the Methodist Story.* Nashville: Abington Press.
Dryfoos, J. G. (1994). *Full-service schools: A revolution in health and social services for children, youth, and families.* New York: Jossey-Bass.
Dryfoos, J. G., Quinn, J., & Barkin, C. (2005). *Community schools in action: Lessons from a decade of practice.* Oxford: Oxford University Press.
Garnett, C. V. (2008). *Interview.* Washington, DC: J. W. Richardson.
Hungerford, B. (February 11, 1974). *Anita T. Anderson: Coatesville Hall of Fame.* Coatesville, PA: Chester County Historical Society Archives, Coatesville Record.
Jago, A. G. (1982). Leadership: Perspectives in theory and research. *Management Science, 28*(3), 315–336.
Johnson, P. (2006). *Interview.* Coatesville, PA: J.W. Richardson.
Kelly, L. C. (2008). *Interview.* Oakland, CA: J.W. Richardson.
Kouzes, J. M., & Posner, B. Z. (1992). *The Leadership Practices Inventory (LPI).* San Francisco: Pfeiffer & Co.
Kouzes, J. M., & Posner, B. Z. (1995). *The leadership challenge: How to keep getting extraordinary things done in organizations.* San Francisco: Jossey-Bass.
Popper, M. (2000). The Development of Charismatic Leaders. *Political Psychology, 21*(4), 729–744.
Rice, D., & Harris, M. M. (2003). Leadership in community schools: A frame analysis. *Reclaiming Children and Youth, 11*(4), 216–221.
Riehl, C. J. (2000). The principal's role in creating inclusive schools for diverse students: A review of normative, empirical, and critical literature on the practice of educational administration. *Review of Educational Research, 70*(1), 55–81.
Sanders, M. G. (2001). The role of "Community" in comprehensive school, family, and community partnership programs. *The Elementary School Journal, 102*(1), 19–34.
Turpeau Sr., D. D. (Unknown). *Up from the cane-brakes: An autobiography.* Cincinnati: Self-published.
Walker, V. S. (1996). *Their highest potential: An African American school community in the segregated south.* Chapel Hill: The University of North Carolina Press.
White, J. A., & Prince, I. A. (2007). *Interview.* Silver Spring, MD: J. W. Richardson.
Willis, D. (Ed.). (1992). *Early Black photographers 1840-1940.* New York: Schomburg Center for Research in Black Culture.
Young, A. S. (1963). *Negro firsts in sports.* Chicago: Johnson Publishing Co.

# 5

# James Adams Community School: Manifesting the Vision

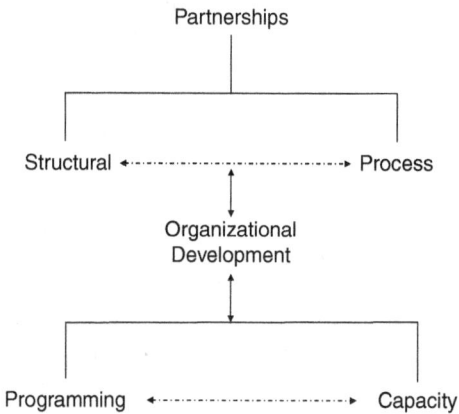

Keep On Going
Life ain't been easy, but I don't complain
I'll take another lick at life just the same.
Most time life sticks me like a brier.
I keep on reaching higher and higher.
(Anita Anderson)

From James Adams' inception, a philosophy prevailed that a school's responsibilities extended beyond the school day. During each of their tenures principals reached out to the East End community in their own way. In 1935, one such example was the pledged fund of $500 to start a recreation center in the East End (Chester County Historical Society Archives, June 4, 1935). It was the Andersons, however, who have been credited with promoting the school – community interdependence that blossomed into the organization examined here. Mr. Paul Johnson confirmed this assertion when being interviewed about displaying of the school bell as a historic landmark commented, "it [JACS] was more than a school. It was used all the time. At night it served as a community center. There were evening classes, dances, plays. The library was open; [and] people would be studying" (Davis, 2005). Mr. Anderson had the distinction of being principal of the school for the longest period when adding his first and second appointments and for serving the longest consecutive period. Functioning at the helm of an institution for a sustained period provided him with the advantage of time to guide the evolution of the school in the direction he envisioned.

While the seed for building a community school may have been planted by John Dewey and watered by Mr. Anderson's own sensibilities, a more systematic strategy for implementation came when T. J. was awarded a fellowship in 1948 and he spent time at the University of Nottingham. When he returned he began patterning JACS after Village College, a school he had visited in Impington, England. Village College was chiefly a community center with a school as its secondary service, whereas JACS was primarily a school with a community center as its secondary service (Chester County Historical Society Archives, October 19, 1948).

By contemporary standards, the community school organization and programming should reflect the priorities that bubble up from the community itself (Britton & Britton, 1970). These principles are not only consistent with the aforementioned authors, but also Dewey, Clapp, and modern experts on this model of schooling (Blank, Melaville, & Shah, 2003; Clapp, 1971; Dryfoos, 1994; Dryfoos & Maguire, 2002; Kronick, 2005). A common place to begin an institutional assessment is determining the resources available to support a full-service school.

Community asset mapping is a term that refers to a formal assessment of the organizations, individuals, and programs functioning in a service district. As per the community school literature, optimally a list of potential school partners emerge from said assets. Equally as important to the "who" are the ways partners are engaged with the school; how consensus is reached; how power is shared; and what collaborations are intended to accomplish (Blank & Berg, 2006; Blank et al., 2003; Dryfoos, Quinn, & Barkin, 2005; Keith, 1999; Kronick, 2005).

Highly effective community schools as per the HECS model require not only stellar leadership, but also school organizational development and attendant partnerships that support the mission of the school. Here we unearth the ways in which partnerships functioned and examples of the organization's capacity.

## Partnership

Community school partnerships optimally build on existing knowledge and expertise of individuals who have a vested interest in the success of the school and community (Britton & Britton, 1970). Power sharing is an element of partnership development often overlooked. Delegation of authority may take many forms, one of which, specific to principals, is protecting the creative space of teachers as much as possible from edicts that might thwart their effectiveness.

The best place to begin reconstructing the types and nature of JACS partnerships is where T. J. and Anita Anderson began, at the neighborhood level. Partnerships in their minds were an exercise in matching interests, skills, and passions in order to ultimately create goodwill and networks of reciprocal support. Ultimately partners aid in delivering a comprehensive educational experience for students (Chester County Historical Society Archives, December 1953). Stated another way, the right people needed to be in place to move the school in a progressive and innovative direction.

Partners with JACS can be divided into general categories: the faith community, Black professionals, White professionals, Black business leaders, Whites representing the business community, persons of national notoriety, scholars, parents, members of the Black community, political actors, and teachers and school staff. How they were identified and recruited is not clear in all cases; however, given the documentation, it is reasonable to assume that the Andersons' charisma and their astuteness in identifying points of mutual benefit between the school and prospective partners contributed to successful engagement. Alignment of the vision of the school with the fundamental mores of the Coatesville Black community added to the attractiveness of potential collaborators, as well as successful student outcomes.

Mr. Anderson's sensibilities about partnering are supported in the resilience and community school literature. Identities of partners are structural and how they are engaged and what they do is process. Though in Table 5.1, partnership structural and process attributes have been separated for the sake of clarity, one cannot sustain structural or process effectiveness in isolation from each other.

Over time, it seems that "belonging" to the JACS community bred both a desire to expand the scope of participation (process) as well as diversifying partners supportive of school programs (organizational development).

Parents, professionals, entrepreneurs, and skilled and unskilled laborers of African descent primarily lived in two sections of Coatesville: East End, where JACS was located, and South Coatesville. Residential segregation

**Table 5.1** Partnership Comparison

| Partnerships | Best Practices | JACS |
|---|---|---|
| Structural (Who) | Inclusive of community assets and organizations Businesses Community persons Parents Teachers and staff Political actors Service agencies | Teachers Parents Local business Supportive superintendents Faith community Civic clubs Colleges and universities Medical community Advocacy organizations |
| Process (What & how) | Opportunities for decision making Meaningful Roles for youth Creates opportunities for creative expression Engages marginalized groups Mentoring Tutoring Job shadowing Supply provision Joint Planning Shifts in professional roles | Community-generated classes Faculty-initiated classes and clubs Shared leadership Delegation of responsibilities Shared vision and ethos Partners saw evidence of their contribution Transparency Demonstrative for support Home visitation Patrons Volunteer and paid staff |

*Source*: Dryfoos, 2003; Harris & Hoover, 2003; Keith, 1999; Sanders, 2001.

had in some respects advantages such as proximity. The lot of community members, regardless of economic station, was inextricably linked. Another benefit of proximity was the concentration of craftsmanship, intellectual, and fiscal resources with sufficient numbers of roles anyone could assume. As an example, the homeroom mothers had the official role of coordinating hygiene checks, identifying needs of students to the teacher and vice versa.

The active parent – teacher association (PTA) conducted fund-raisers, which supported both the short- and long-term aspirations of students. Upon graduation from JACS, PTA-sponsored scholarships in various categories were awarded. Pressing needs such as raincoat sets for the safety patrol and underwriting the cost of annual field trips to the Pomeroy Farm and other locales were also beneficiaries of their support. PTA funds also filled a gap for children who qualified for the federally subsidized free and reduced lunch program. Though the cost of the hot lunch was reduced with governmental funding, some families still did not have enough money to pay even the lowered fee. As a result, the PTA provided subsidies. Another example of the effective partnership between the PTA and the school was the reopening of the library after a fire in the early 1940s. Under Mr. Anderson's leadership, the library, reopened in 1944 with over 2000 books and was dedicated to the first Coatesville young man who was killed in military service, James Boyd. Borrowing privileges were extended as one might predict to anyone in the community (Chester County Historical Society Archives, May 17, 1944a, May 17, 1944b, November 1, 1946).

Civic organizations such as Jack and Jill also subsidized scholarships awarded at graduations ceremonies. Jack and Jill of America, Inc. is an organization founded in 1938 in Philadelphia that supported the educational attainment of African American children through cultural, civic, recreational, service outreach, and political advocacy. The Young Women's Civic Club of South Coatesville, The American Legion, and the Dansforth Foundation were also sponsors of scholarships and service awards. Youth organizations that functioned in concert with their parent agencies, such as the Junior NAACP, worked as partners supporting the activities and mission of the school (Chester County Historical Society Archives, January 29, 1944).

Much attention is paid to the role of teachers as leaders in this community school because without their dedication to the vision implementation would never have taken place. Mr. Anderson commented that the conversion of James Adams to a true community school would not have been possible without the commitment and stellar credentials of the teachers. His exact words were, "Because of the superior skills and attitudes of teachers, it has been possible not only to greatly expand and adjust the curriculum to meet changing conditions, but also to enrich the total school program" (T. J. Anderson, 1950, p. 9). One of the roles of the organizational leader is to ensure teachers have the space to function optimally. In the words of T. J., "Without school administrators who accept its [JACS's] basic philosophy and are willing to run the interference for those who are carrying the ball (teachers), no school staff can proceed very far" (T. J. Anderson, 1950, p. 9). Thus from T. J.'s perspective a school principal, faculty, and staff must not only commit to the ideals espoused by the community school, but they each need to support each other as partners and share power to ensure the work at hand is accomplished.

The teachers at JACS were quite accomplished by any standards. For example, William H. Jones, in 1949, was the president of the Coatesville

Teacher's Association. This graduate of both Virginia Union University and Columbia University was known for aiding students desirous of attending college. At a testimonial in his honor it was said that he helped more students attend college on a shoestring than anyone and "if the student lacks even the shoestring, Mr. Jones finds that shoestring somehow—even sometimes from his own pocket" (Chester County Historical Society Archives, February 19, 1952). Leonard Dorsey (a graduate of West Chester State Teachers College and the University of Pennsylvania) wrote two penmanship books, held a patent for a ruler that corrected penmanship, and submitted several other patents, one for a radio tuning dial. He was also listed in the 1956 edition of "Who's Who in American Education" (Chester County Historical Society Archives, June 6, 1956, June 17, 1954).

Paul Palmer with Howard and Columbia University credentials oversaw much of the recreational programming in Coatesville and in addition was the Executive Secretary of the World War II Memorial Community Center and was the President of the Chester County Chapter of the Pennsylvania State Education Association (Chester County Historical Society Archives, December 13, 1965). William H. "Pop" Ransom was a graduate of Lincoln University (of Pennsylvania) and Temple University, having also matriculated for a time at the University of Pennsylvania. He oversaw the organized sports teams of the JACS community center and taught science. Accolades as a coach were reflected in the individuals who after his tutelage went on to play professional sports, participate in the Olympics, and play on college teams. Under "Pop" Ransom's leadership, during the first 3 years of JACS's independent track team (separating from Gordon Junior High), his athletes won the Chesmont League Track Junior High Championship and ran in the Penn Relays in Philadelphia (Chester County Historical Society Archives, February 10, 1954, May 1954).

There were many women teachers at JACS, who were also well credentialed, such as Mrs. Edna Keith, Priscilla Elizabeth "Betty" Cuff (later Cuff-Stokes), Eliza Howard, and others. Unfortunately, there are virtually no accounts of their achievements in the local press. What we know, however, from the many pictures of school activities and who ran them, is that women had prominent leadership roles in the school. Also, the fact that JACS had the best-credentialed staff in Coatesville implies that the female faculty members were as likely as the males to possess master's degrees. One verification of this assumption was found in a 1953 school board report announcing that Mary Jane Moore earned her M.A. from Columbia University and was selected for an audition for the Marian Anderson Award (Chester County Historical Society Archives, October 1953).

One student summed the influence of teachers in one phrase, "we were groomed for excellence." T. J. concurred with students and was known to write that the versatility, community-minded zeal, boundless energy, and professional preparation of his teachers were the foundation for JACS's success (Chester County Historical Society Archives, June 1952). Teachers were so committed to their students' well-being and educational attainment, that even after students moved up to the integrated high school, they would make sure students were taking courses representative of progression from the JACS coursework. Several alumni remembered teachers who took them to visit colleges if they expressed an interest in matriculating. As a testament to their being held in high esteem even in the White community, it is remarkable to note that every one of the JACS teachers was offered jobs in Coatesville area schools as they were gradually desegregated as a result of the *Brown v. Board of Education*. This was not the case in many places across the country where Black teachers were displaced on the heels of desegregated schools (Morris, 2008).

As teachers were viewed as partners in community school endeavors, their special talents and interests informed the program offerings. As Ms. Cuff-Stokes, who was both a student at Adams and then a teacher, recounts, it was no imposition for teachers to return in the evening to help students or work with clubs and other activities. She reminisced, "we were happy to do it" (Cuff-Stokes, 2007). While clearly by today's standards the time commitment would seem unreasonable, because the teachers were part of the community and were committed to its uplifting, the lines between their professional and personal lives were as blurred as the Andersons. This commitment also extended beyond time into personal resources, as many scholarships, awards, and program patrons were teachers.

One strategy imposed early in the Anderson administration as a means of bridge building between the school and families was home visitation. The expectation was articulated in a report to the school board:

> Teacher visitation at Adams is an integral part of the homeroom guidance program. Each teacher is expected to know each homeroom pupil's parents. The face to face contacts with the adult members of the family in the place where the child is accepted without reservations is a vital aspect of the school's public relations program. Home and school become genuine co-partners in the great enterprise of making democracy work. (Chester County Historical Society Archives, December 1953, p. 2).

Anderson clearly was convinced that building goodwill between parents and teachers would accrue not only to students, but also improve the quality of school life holistically. The quote also confirms the notion that parents and teachers were necessarily partners in the learning – teaching exchange.

Facilities at JACS often served multiple purposes that sometimes required the rapid shifting of furniture and alternative setups. Espousing the value of home visitation in another document, Mr. Anderson reported to the school board an occasion when distinguished guests were due to arrive for a program shortly after the school day. When he arrived to begin moving furniture and rearrange the room for the reception, parents had already completed the task. He was convinced that the goodwill created between teachers and parents had been fostered by the home visitation, and contributed to a solid foundation of trust between the school and broader community (T. J. Anderson, December, 1953; Chester County Historical Society Archives, December 1953).

Staff (to include a secretary, dietician, and custodial personnel) also played a pivotal role in the life of the school. Respondents still remember with fondness the culinary giftedness of Mrs. Gibney, the school dietician. She not only cooked for school lunches, but also prepared food for school-sponsored banquets, teas, and meals for other organizations hosting events at the school. Staff also participated in other sectors of after-school programming. For example, there were always contests, such as the Evelyn Lawrence Oratorical Contest, in need of judges and plays in need of stage assistance and performers (Adams Community School, February 5, 1954). It was noted by several respondents that Mr. Anderson treated everyone as a valued contributor no matter what their professional station.

Students of JACS were not only the recipients of instruction and activities; they also contributed to the vibrancy of the community. Furniture made in shop classes were donated or sold to support the school. Girls and boys raised money to donate to local and national charities. The safety patrol instilled a sense of responsibility in the older children for the younger ones. JACS was a school where the majority of students walked and there was a need to ensure that particularly early elementary students would arrive and depart safely. Pupils in upper grades who had demonstrated their discipline as good students could be nominated by teachers or others to join the safety patrol. Patrol members stood on corners and directed the passage across heavily traversed streets. Participation in guarding younger students instilled a sense of responsibility, caring, and belonging to the community that carried over in and out of school. I know first-hand how this could be so as in my time the honor of serving as a safety patrol in this same town was bestowed upon me.

Churches and religious organizations such as the YMCA were instrumental in supporting programming as well. For years the local YMCA supported the after-school programming with money and volunteers (*Adams Gazette*, June 1952). Pastors from numerous faith traditions were invited to provide the invocation or benediction at school functions, some of which were not Negro or Christian. As examples, rabbis from nearby synagogues and pastors of Holiness, Baptist, Seventh Day Adventist, Methodist, Presbyterian, and Catholic churches were regularly part of programming. Some of these churches had JACS children and families in their parishes and some did not. Despite that, there seemed to be a dependable flow of support from the Coatesville and surrounding area clergy for the work transpiring at JACS irrespective of race.

Partners did not have to be limited to persons living and working in the surrounding community. The JACS partnership web extended beyond Coatesville to engage persons who could inform, inspire, and enlighten students and adults. Tapping largely into the Andersons' influence and network of family and college friends, scholars and persons of international fame were sometimes hosted by virtue of cosponsorships and at other times for specific JACS programs. Whenever these events took place attendance was open to the general public. Paul Robeson (actor, athlete, author, civil rights activist, basso cantante concert artist), Leontyne Price (renown soprano with the Metropolitan Opera Company of New York), Philippa Schuyler (a child piano prodigy), Roland Hayes (concert vocalist), all performed in Coatesville at the behest of the Andersons. Activists such as Dr. Lester Granger, Executive Secretary of the National Urban League, Anita's mother Ila Turpeau, and scholars from historically Black colleges and universities (HBCUs) frequented JACS. Dr. Leslie Pinkney-Hill, President of Cheyney Teacher's College, Dean William Menchan, also of Cheyney, William Gilmore of Norwalk, Connecticut, and Charles Huston, Jr., President of Lukens Steel Company, spoke at graduation ceremonies. Given the affinity of Coatesville residents for sports, it was quite a coup to secure a visit from Wally Triplett, an all-round Penn State athlete and the first player to break racial barriers in the Sugar Bowl played in Texas in 1949, who went on the play for the Detroit Lions (Adams Community School, June 4, 1954, June 5, 1951, June 7, 1949, June 8, 1948; Chester County Historical Society Archives, December 31, 1945, February 23, 1944, May 28, 1955).

Many of the aforementioned partners shared time and expertise. However, volunteerism alone tends not to be enough to sustain a highly functioning community school. The predominant evidence of fiscal engagement with JACS was found in the patronage of the yearbook, scholarships, and service awards, all coming from diverse sources. One of the skill sets T. J. brought to JACS was that of a business and public relations manager preparing him well for his tenure in Coatesville. Fiscal support from White partners was impressive and included the National Bank of Chester Valley, Lukens Employees Cooperative Store, Sante's Pharmacy, East End Hardware Store, Rice & Weidman (building contractors), and

M. Chertok (furniture store). Though not an exhaustive list, they were the most frequently noted patrons of programs and scholarships. Black businesses also supported the school, such as the Clement Atkinson Memorial Hospital, Prout Funeral Home, and Lincoln University. The longer the Andersons worked at JACS, the lengthier became the list of contributors and the more diverse were the sources of support.

As one might expect, some partners were more vested than others. Lukens Steel Mill and its affiliates such as the employee cooperative store, and the United Steel Workers Union were visible more than once annually. Lukens published an employee magazine called "Lukens Life" and in it each month happenings at local schools (to include JACS) were mentioned. Charles H. Huston, Jr., President of Lukens Steel Company, was the keynote speaker at JACS's graduation ceremony in 1951.

Dr. Whittier Clement Atkinson was a dependable advocate for the health of students and the community and provided practical assistance in the way of medical services on-site. Another medical partner was Dr. Thomas Clayton, who was employed by the school board in the early 1950s.

Political advocacy was one way to demonstrate solidarity with the school and for our purposes represented a type of partnership. Earlier we considered the value of political allies such as superintendents Vanderslice and Rice. There were others as well. In a letter to the Coatesville School Board, Dr. Clayton resigned his position in 1955, because of the board's refusal to integrate schools. He notes,

> I must send in my resignation because, although, as I have said above, I feel highly honored to serve in the capacity of such a position, I can no longer do so in view of the fact that the School Board has not seen fit to integrate. My present stand is made urgent by the fact that integration should take place both in the light of the recent historic Supreme Court Decision and also the northern tradition of integrated schools. (Chester County Historical Society Archives, October 22, 1955)

As a commentary on his loyalty to JACS, he remained available to treat students on an emergency basis for no charge. Of all of the outspoken supporters of JACS one of the most visible and politically influential was James Brown. Rev. Brown was the pastor of the Upper Octorara Presbyterian Church, was a close friend of T. J.'s and he gained national attention as a leader in the charismatic movement. Long before meeting the Andersons he demonstrated his disdain of segregation by participating in protests and even being jailed on at least one occasion in Washington, DC (Chester County Historical Society Archives, October 17, 1946, October 30, 1946).

Rev. Brown and others like him are examples of the power of political actor-partners in garnering ongoing support for a community school. Dimensions of leadership discussed earlier resurface as we consider partnerships because a willingness to share power is paramount when setting the tone for creativity in the organization and imbuing partners with the authority to implement ideas. The program offerings after school were not identical each year because community members were asked to articulate desired courses/activities that could be offered in addition to the programming advanced by the school faculty and staff. Soliciting suggestions that are then implemented are one way power was shared with the community and afforded opportunities to influence the life of JACS even if individuals did not have children in attendance.

There were times when Mr. Anderson was authoritative, such as when his teachers went to citywide meetings where he instructed them not to sit with each other, but rather mingle with others to learn what was being done at other schools. He was also the authoritarian when it came to misbehaving students, as affirmed by one who wrote, "He remembered you day one. If he had to see you in his office, he would follow-up to make sure you were on track." Yet, he was not consumed with a need to control everything as exhibited by the confidence he had in his staff. Much like the heads of contemporary organizations, he seems to have spent a great deal of time being the face of JACS in the larger community in an effort to advance equitable educational opportunities for students and the cause of racial harmony.

Harkening back to the HECS model, effective partnerships posses both structural and process attributes. Friends, political allies, parents, adults without children in the school, faculty and staff, the business and faith communities were all potential collaborators. A willingness to contribute was all that was required, which provided a diverse set of skills and interests to create the foundation for the community effort. There are no hints in publications or interviews that these tendencies were viewed as opportunistic by students, faculty, or members of the general community. Instead, there seems to have been a perpetual flow of ideas and resources between the school, its service neighborhood and beyond.

Tentacles of service extended beyond school-based initiatives and there were always CIHU opportunities for partners. As one instance, teachers of Adams in 1954 contributed $1195 collectively to support the expansion of Atkinson Hospital (Chester County Historical Society Archives, October 1954). Harvest Home, a JACS program designed to collect groceries for the Atkinson Hospital, collected five bushels of food one year (Chester County Historical Society Archives, December 1953). During Halloween trick-or-treating, pupils collected 53 pounds of soap, thread, and needles to be sent through the American Friends Committee (Quakers) to needy Korean families (Chester County Historical Society Archives, November 1954). Joining forces with other organizations the JACS community aided in collecting funds to create the World War II Memorial Center (Chester County Historical Society Archives, June 6, 1945).

Another secret to the depth and breadth of JACS collaborators was the penchant to publicly acknowledge gratitude. In the absence of reciprocity, sustainable partnerships are improbable. Thus banquets and assemblies were regularly devoted to honoring those dedicated to JACS and the elevation of the Negro condition in Coatesville. Steadfast collaborators were recognized with annual awards such as—the Americanism Award and the Citizen Cup.

Structural dimensions of partnerships speak to who the partners are while the process of partnership speaks to what they do, how partners influence the functioning of the community school, and how power is shared. Garnering allies and sustaining said relationships over a period of 13 years was in part attributed to the flexible leadership style of Mr. Anderson. He was a savvy and authoritative leader when he needed to be, and simultaneously willing to provide space for partners to make decisions best left to their expertise. Distributed authority also empowered teachers, staff, and community members to identify partnership opportunities. I speculate that what appeared to be an ease in releasing control was informed by the broad consensus about JACS's mission and purpose. Collaboration may take many forms. There were more than enough venues for episodic or sustained engagement with the school community. Of those who remained wedded to the work of the school it is clear they mirrored what the contemporary literature cites as most critical to building sustainable community schools (Blank & Berg, 2006; Dryfoos, 2003; Keith, 1999).

Yet all of this enthusiasm could not have been sustainable in the absence of an organization with the capacity to showcase efforts. As we move from the nature of partnerships HECS model affirms aspirations of a community school cannot be realized in the absence of careful attention to the structural and process factors that contribute to efficient operations or organizational development.

## Organizational Development

As per the model driving this analysis (HECS), a school organization can be divided into two principal categories of programming and capacity.

Table 5.2 provides a glimpse of organizational programming and capacity as espoused by researchers and demonstrated by accounts of JACS. A community school organization arranges itself to perpetuate programming using fiscal and in-kind support. Sectors supplying aid to the community school, as was noted in the previous section, can provide a

**Table 5.2** Organizational Development and JACS

| Organizational Development | Best Practices | JACS |
|---|---|---|
| Programming | Qualified teachers<br>Curricula with clear performance expectations<br>Programming linking academic content to life experiences | Operettas<br>National Junior Honor Society<br>Christmas cantatas<br>Achievement banquets<br>Dramatic club<br>Testimonial receptions<br>Evelyn Lawrence Oratorical Contest<br>Dance recitals<br>May Day<br>Adams school bank<br>Move-Up Day<br>Class Day<br>Annual Negro History Week<br>Winter carnival<br>Literacy classes<br>Game nights<br>Intramural sports<br>Spring musicals and plays<br>Discussion groups<br>Income tax clinics<br>Citizenship classes<br>Arts and crafts (woodworking, leather work, ceramics, toy making)<br>Boy and Girl Scouts<br>Mother–daughter banquets<br>Orchestra<br>Piano lessons<br>CIHU cupboard (clothes closet) |

**Table 5.2** (*Continued*)

| Organizational Development | Best Practices | JACS |
|---|---|---|
| Capacity | Extended day activities<br>Family centers<br>Health clinics<br>Adult evening classes<br>Quality facilities<br>Adaptable facilities<br>Financial support<br>In-kind donations<br>Skilled volunteers | Volunteers<br>Teacher-volunteers<br>YMCA funding<br>School board fiscal support<br>Program patrons (businesses, faculty, staff, community)<br>Medical volunteers<br>PTA donations of time and money<br>Facility expansion<br>Facility convertible from day to evening activities |

*Source*: Dryfoos, 2003; Harris & Hoover, 2003; Keith, 1999; Sanders, 2001.

wide range of support. Organizational capacity takes into account how programming is managed and resourced in tangible and intangible ways. Our attention here is on the specific activities that took place on- and off-site and the revenue sources in addition to the school board – allotted budget that made programming possible.

### Programming

Activity around JACS even in retrospect seems dizzying and a challenge to reconstruct. With very different age groups housed in such close proximity, one might anticipate problems. However, there seemed to be little evidence of friction between grades or a need to "protect" the younger students from the older pupils. Instead, whether a function of the pervasive implementation of CIHU or some other forces, students seemed to go about their course work and activities with little time for anything else. Adding to the stability of the environment were teachers who not only knew students and their families, but also interacted with pupils outside of school as most lived in the same neighborhoods as the students.

During the school day elementary students studied the basic subjects of penmanship, mathematics, English, social studies, and science. As students ascended in grades, more course work options were available. Language requirements for the seventh grade were Latin and German, and eighth-graders took French and Spanish. Exploring these languages exposed student to rudiments that informed their high school course selection and laid the foundation for vocabulary prowess (particularly as it pertains to Latin). Evidence of CIHU's infiltration into classrooms and extracurricular activities could be found around every corner. For example, home economics and shop classes were available in upper grades and projects were often donated to charities or to the school itself. One example was the shop classes' involvement in the reconstruction of the school library after the fire (Chester County Historical Society Archives, May 17, 1944b).

Academic achievement was stressed as a means of personal betterment as well as a vehicle to enhance the type of service one could render to one's community (local, national, and global). Alice Carter (a former student) reflected, "We received a special blessing at the time in that we had some of the most brilliant teachers, because they weren't allowed to teach in other schools, but they still poured all their knowledge into the students at James Adams. They taught us how to be the best we could be" (Robinson, 2005). In geography and history classes more than facts were taught. Youngsters were cajoled to view themselves as a citizen of the world with all the rights and privileges that that designation afforded.

Realities of segregation were not ignored, nor the need to advocate for change. The school culture managed to strike an affirming balance between social and opportunity limitations of the present and what could be achieved in spite of barriers. Students were clear that they "had to be twice as good to get half as far as Whites," and were encouraged to build the foundation for excellence first by mastering academic disciplines. Interestingly at the same time bitterness was not tolerated and examples of progressive Whites regularly formed part of school life, which kept students from ascribing narrow definitions to entire races of people. Harmony was struck between the imposed limitations of living in a segregated city and the need to advocate for social justice. Too much attention to either of these realities might promote anger, frustration, or hopelessness. This tension was mitigated with the hope that defying predictions of mediocrity and demonstrative benefits of JACS's contributions to the larger Coatesville community would create unprecedented opportunities for students. As a result, every club, activity, play, and school assembly was devoted overtly or covertly to sustaining that delicate balance.

Curriculum was brought to life with clubs, oratorical contests, cantatas, operettas, dramatic presentations, and concerts of many kinds. Extracurricular activities in many cases were laboratories for classroom content (Chester County Historical Society Archives, November 1953). During the Andersons' first stint in Coatesville, Anita established an annual oratorical contest. Its name was changed to the Evelyn Lawrence Oratorical Contest after the seventh-grader who died shortly after being named the initial winner (Chester County Historical Society Archives, February 1954b). Texts to be recited as part of the contests were a mix of the African American and Eurocentric canon, for example, Langston Hughes, James Weldon Johnson, Shakespeare, and Homer. Selections reinforced English/language arts and history lessons. Geography lessons were embedded in traveling exhibits brought to the school such as the toys from elementary students in Darmstadt, Germany (Chester County Historical Society Archives, February 19, 1949).

Science jumped from the textbook pages when, for example, Miss Johnson's fifth grade class's prehistoric animal exhibit was selected to participate in the science fair held at the well-respected Franklin Institute in Philadelphia (Chester County Historical Society Archives, April 1954). Annual trips to the Pomeroy Farm, where students collected specimens and brought them back to the school for further examination, made botany and biology tangible (Chester County Historical Society Archives, May 27, 1952).

"Think Twice Before You Leap" was a demonstration lesson of an eighth grade JACS math class in the window of a large White furniture store (Chertok's Furniture Store) during that year's American Education Week. The purpose of the lesson was to investigate the cost of paying for items using installment plans. Students concluded after their computations that retailers would prefer to sell for cash and that installment purchases exact additional costs on both retailers and purchasers. As a result, one should "think twice" before assuming an installment loan (Chester County Historical Society Archives, December 1954). The Adams Bank pitted classes and grades against each other in savings contests. The winners were determined not so much by virtue of actual funds raised as the percentage of classroom students participating (*The Adamite*, April 1949). In both the Chertok and banking example, the relevance of acquiring and applying mathematical skills was emphasized to students in ways they could immediately identify as relevant.

Physical fitness and its relationship to long-term health was the premise reinforced in the annual May Day celebrations. Each May, students would prepare dances from different parts of the world to perform. Festivities included a parade, drills, gymnastics, and the election of the May Day Queen and King (Chester County Historical Society Archives, May 1954). In addition to May Day, themes promoting general health were integrated into science, home economics, and physical education curricula. Advancing positive health outcomes was deemed paramount in the nurturing of holistic development and a happy and productive life (Chester County

Historical Society Archives, February 1954a). In other cases activities were designed to foster artistic competencies or just fun.

Students were afforded opportunities to participate in operettas (what we now refer to as musicals), combining public speaking skills, musical talent, and often dancing. Banquets of all sorts were held where proper etiquette was rehearsed. In some cases the young ladies in home economics classes would cook and their male classmates would come eat, practicing table manners. Some assemblies were student led, such as the "Three R's Hour" hosted on the fictitious CIHU station supported by the Adams Broadcasting Network. Ninth-graders using a typical radio format sought to "sell" after school activities to the student body. The show was complete with commercials and the playing of popular music (Chester County Historical Society Archives, March 1954). Broadening exposure even more were opportunities for students to perform on actual radio stations, such as Coatesville's WCOJ. One gets a sense, when rummaging through the archives of school programs and pamphlets, that programming bridged gaps in student's exposure and sought to address them in ways that did not highlight deficits. (Chester County Historical Society Archives, April 1954, October 10, 1945).

Guidance clinics were also used to provide a different type of support. Meetings were established for graduating students with persons in their respective interest areas, such as vocational training or medicine to ensure they knew the correct courses to take once they matriculated in the integrated high school. Once a JACS student, the resources of faculty, administration, and staff remained at your disposal. It was common that aid with college applications and even support after matriculation would come from JACS. With this level of support, African American students in Coatesville went from a handful graduating over a period of 10 years to students attending Ivy League, Big Ten, state colleges, and HBCUs in large numbers (Chester County Historical Society Archives, December 1954).

A merit system established a way students could accrue points for extra accomplishments. Children in the upper range of points at the end of the school year received not only awards, but also guest seating in the Annual Achievement Banquet where they were publicly acknowledged (Chester County Historical Society Archives, April 1954). This was a way to affirm students whose gifts might be overlooked if a focus was solely on academic performance.

Students, faculty, and community members were acknowledged in a plethora of forums. Some of the awards conferred at the annual achievement banquet included the Americanism, Citizenship and Alumni cups, Fellowship Citations, and Student Oscars. The Americanism Cup was awarded to a local citizen who had made a significant contribution to the problem of making democracy work in Coatesville. The Citizenship Cup was given to an East End resident who had raised the cultural level of the Adams School Community. The Alumni Cup was conferred on an Adams alumnus for his or her distinguished achievement during the current decade. Fellowship Citations were for citizens whose lives reflected the CIHU spirit.

Student Oscars were categorized into educational levels. The Graduate and College Oscars went to Adams alumni with the most outstanding records in graduate and undergraduate studies. Similar awards were conferred on Adams alumni in high school. Current students with the highest scores on the merit point system, best all-round male and female athletes, and excellence in arts and crafts, music, dance, and dramatics also earned these Oscars. Finally there was an Oscar given to the most valuable boy or girl in the community center athletic teams (Adams Community School, May 25, 1956).

Faculty were not always working as sponsors and mentors; they also had a chance to reveal their fun sides. Though teachers were held in the highest regard by students, parents, and the community, this penchant for pure entertainment seemed not to damage their credibility in the classroom. Instead it made them seem more accessible and more attainable role models in the eyes of their students. Spring musicals and plays joined students and faculty as equals on stage. Take for example the Adams production "A Night in Spain," where several teachers played beggars or friends in the script and students danced, sang, and filled out the cast (Adams Community School, April 29, 1955). Numerous other examples exist, such as the assembly where students imitated Mr. Anderson and teachers while teachers acted as students in the audience.

Move-Up Day was another celebration where teachers publicly rewarded students who had performed well academically and were to be promoted to the next grade. Like other Adams programs, there were songs, dramatic selections, awards conferred, and then the presentation of promotion cards by Mr. Anderson with much pomp and circumstance (Adams Community School, June 13, 1956).

Great emphasis was placed on the fine arts, and opportunities for students to identify their talents and hone them were numerous. For example, one graduate went on to become an anchorwoman for a Chicago news broadcast (J. Diana Robinson Lewis). She credits Anita Anderson for preparing her well for her chosen career through the oratory contests, the use of proper grammar, and practice of extemporaneous speaking, all contributing to the comfort Diana possessed in front of the camera.

Alumni noted how teachers prepared them not only relative to book knowledge, but also how to excel in diverse social settings. Students accounted how comforting it was to know that leaving the "nest" did not mean leaving support behind. They remained connected to JACS in that faculty members were available to advise and advocate on their behalf. More than one student confirmed that faculty ensured students were assigned to the correct classes at the correct levels and made sure those interested in going to college visited campuses and submitted requisite paperwork.

Given graduation from JACS equated to ninth grade completion, it is interesting that at this time scholarships were awarded. The sources of scholarships were diverse, and from newspaper accounts it emerges that the longer T. J. and Anita Anderson worked, the more scholarships became available. Sources of awards came from White and Black organizations, personal sources, and the PTA. The local branch of the American Legion, the South Coatesville Civic League, the Young Women's Progressive Club, and the *Reader's Digest* all sponsored scholarships. The Andersons also sponsored several scholarships; the D. D. Turpeau Award (named for Anita's father established to honor a student who best combined ability, achievement, and courage) and the CIHU Award (for the student who best demonstrated a sustained commitment to community service) are two such examples.

Topics of national and international significance were the subject of student graduation panels augmented by guest speakers. Table 5.3 shares a few graduation foci that frankly have retained their significance.

Table 5.3 JACS Graduation Themes and Speakers

| Year | Graduation Theme/Panel | Guest Speaker |
| --- | --- | --- |
| 1948 | Adams in the Coatesville of Tomorrow | Dr. Leslie Pinkney-Hill, President of Cheyney State Teachers College Rev. Clyde Wilson |
| 1949 | To Secure These Rights | Ila Turpeau, Civic leader Rev. Thomas Pugh |
| 1951 | The United Nations and the Universal Declaration of Human Rights | Charles H. Huston, Jr., President of Lukens Steel Company Father Albion Ferrell |

**Table 5.3** (*Continued*)

| Year | Graduation Theme/Panel | Guest Speaker |
| --- | --- | --- |
| 1952 | Coatesville in a Democratic World | Dr. William Menchan, Dean, Cheyney State Teachers College<br>Rev. James Brown |
| 1953 | The Adams Junior High School Pupil Evaluation | Dr. William C. Brown<br>Rev. George Eggleston |
| 1954 | I Am the Adams Community School | Dr. Lester Granger, Executive Secretary, National Urban League<br>Elder Aaron Brogden |
| 1955 | The United Nations and Human Rights | Cecil Rice<br>Rev. Eggleston |
| 1956 | Democracy and Race | William Gilmore<br>Rev. David Minus, Jr. |

Pupils were not the only beneficiaries of programming. JACS was always open to suggestions from the community for new classes and would create them if enrollment were sufficient. Night classes were advertised in the Coatesville Record and as long as six persons expressed interest, the classes were offered between 7 PM and 10 PM (student programming ended at 9 PM) (Chester County Historical Society Archives, November 13, 1946, October 18, 1950, October 19, 1948, September 23, 1950).

Adult enrichment classes fell into categories similar to those of students, academic enrichment, fine arts, vocational skills, and recreational. For example, adults could take piano lessons, sing with the adult choir, borrow books from the library, or take pottery classes. Recreation for adults included discussion groups, movies, and sports. Furthermore, parents were encouraged to share recreational and educational classes with their children (Chester County Historical Society Archives, October 10, 1945).

There were woodworking and sewing classes (Make-Do and Mend Club), sharpening skills that participants could convert into income-producing endeavors. Three R's refresher courses (Reading wRiting and aRithmetic), GED preparation, assistance in job applications, employment testing, and college applications were provided. As a result, there were adults who after taking evening courses at JACS were accepted to universities, secured GEDs, or successfully completed the civil service examinations

(Chester County Historical Society Archives, October 19, 1948, September 23, 1950). Evening programming was made available free of charge to anyone who wished to come, which brings up the question, Who paid for all of this? The school board did not supplement extensive offerings completely. Thus, each year combinations of fund-raising events such as the winter carnival, volunteers, and teachers provided much-needed support. Organizational capacity speaks to the practical decisions necessary to ensure sustainability in community school programming.

*Capacity*

At their heart, discussions of community school capacity disaggregate the ability of the organization to bring to fruition its mission\vision. JACS's vision was stated in different ways for different audiences but central to all articulations were the priorities of stellar academic programming and outcomes, meaningful engagement with community, and services to first the African American neighborhoods and then to Coatesville at large.

The student body for most of the 1950s exceeded 550 pupils distributed between kindergarten and ninth grade. The smallest grade-level population was about 42 with the highest being 67 (Chester County Historical Society Archives, April 1954, December 1953, December 1954, March 1954, November 1953, November 1954, October 1953, October 1954). Teacher's salaries as well as administrative, janitorial, and office staff were comparable to Whites in Coatesville by the 1950s, as per an example of the city's budget. However, the salaries of the cafeteria dieticians and staff were approximately half that of the Whites in comparable positions. Expenditures for the physical plants seemed also to be distributed across the district based on greatest need, with JACS consistently lagging behind its White counterparts in physical plant investment (Chester County Historical Society Archives, May 24, 1955). In its infancy, the community center aspects of James Adams were supported with three paid staff and 20 volunteers. Over a period of 10 years, the paid staff grew to ten with at least double the initial number of volunteers, which does not include teachers (Chester County Historical Society Archives, March 22, 1954).

Organizational capacity of JACS cannot be determined by city investments alone. Consider a snapshot of the attendance generated in evening events. For the period between October 15, 1953, and March 15, 1954, total attendance at all after-school events was 13,784. This period included

95 sessions, 23 activities sponsored by organizations other than the school, and 363 group meetings that could include organizations affiliated with the school or loosely coupled by virtue of location (Chester County Historical Society Archives, April 1954). Students and T. J.'s reports to the school board confirm that infusions were needed to support the robust evening activity schedule. The major fund-raiser was the winter carnival. Booths selling food and clothes were side-by-side games. Clowns, acrobats, movies, and talent shows were part of the evenings' events. It was not uncommon for between 300 and 400 persons to participate in such activities at JACS on any given evening (Chester County Historical Society Archives, March 1954, March 6, 1954).

There was also evidence that the intense programming reversed the trends that concerned political actors prior to Anderson's arrival. In a 1954 report to the school board, Anderson reported that "police records will show that juvenile delinquency is at a low ebb in this community. When one views the juvenile problem in various cities and neighborhoods, all around us and then considers the rate of juvenile crimes in our own community, one cannot doubt the service this program [JACS] is giving" (Chester County Historical Society Archives, April 1954).

Capacity also begs consideration of the physical plant that was home to all the day and evening activity. The numerous annexes added square footage to the building, yet there never seemed to be quite enough room to support the different types of classes or events. As a result, with the exception of the principal's office, the CIHU cupboard, and pantry (sources of clothing and food for anyone in need), every inch of the school served multiple purposes. Classrooms were converted to game rooms; the gym doubled as a banquet location. The perpetual shifting of usage added another capacity issue that seems to have been addressed by anyone who was nearby, as in the moving of furniture and special staging.

As per the Children's Aid Society, community schools as organizations need to strive for five main goals: extended learning, parental support and involvement, community building, innovative curriculum, and integration of health (The Children's Aid Society, 2001). All of these admirable goals are impossible to achieve in the absence of fiscal, in-kind, and volunteer reserves. Fueling the programming, partnership development, and capacity was a commitment to CIHU. CIHU threads are not easily extracted or retrofit into a singular category in the HECS model because it coursed through every dimension of life at JACS. As a result, a discussion of its relevance to partnership and organizational development merits its own attention.

## Living CIHU

CIHU Song
CIHU CIHU Oh my friend can I help you
(Mary Ryan)

If one acronym embodies the JACS vision, it was Can I Help You (CIHU). While discussed in the context of symbolic leadership, it also has a place as we consider JACS's organizational capacity and programming. Even before CIHU officially was pronounced in 1946, the tenets of this philosophy had already driven the Andersons. In 1955, Dr. Amos Travers, a professor at Hamma Divinity School and advocate for Dr. Laubach's CIHU theology, visited Adams school and reflected, "An indefinable something in the spirit of the school is reflected in the happy faces of students and faculty. It is a community spirit, full of goodwill. Even salesmen of books and equipment feel it; they like to visit Adams whether they make a sale or not" (Chester County Historical Society Archives, April 4, 1955).

More than 50 years after the closing of JACS, students still refer to the spirit of CIHU and how it influenced them. When asked what CIHU meant, they recount two major themes, being recipients of the CIHU spirit and how it became a lifelong philosophy. Samples of remembrances are the "spirit of togetherness was rampant"; "It encouraged me to achieve in my own right but it encouraged me to help others achieve"; "I will maintain that feeling as long as I live. I practice it everyday"; and "It helped me to try to give help as well as accept it." Relative to the CIHU commitment of the faculty, Andersons' daughters Jean and Ila said, "the principal, teachers, and staff demonstrated this caring spirit daily to students and those in the community." Dedication to CIHU and its tangible evidence came to life in the donated boxes of food and clothing to families and Atkinson Hospital; the free music and piano lessons from Mrs. Anderson; the CIHU cupboard and pantry, "trick-or-treat" used to collect supplies for needy families in Korea and many other examples. The intangible evidence of CIHU could be found in the ways adults and children treated one another, actively sought ways to be of service without prompting and how influential this way of viewing one's purpose in the world became. Any doubt of the potential lifelong impact of CIHU dissipates when sitting in a room of JACS alumni, who still can sing the CIHU song with their voices and their lives.

JACS programming and capacity as exemplified in this chapter clearly align with contemporary recommendations. Inclusion of individuals and group entities with a natural vested interest in educational outcomes were systematically engaged. This group alone is not descriptive of the scope of stakeholders. The metaphorical partnership net was cast in such a way that entities who might not automatically have viewed themselves as beneficiaries of positive student and community outcomes came to adjust their viewpoint. Configurations of participants and programming created intergenerational, interracial, and public–private bridges.

In order for programming ideas to be implemented the best organizations demonstrate their ability to distribute resources efficiently. After school classes, clubs and sporting events took place because the facility itself lent itself to multiple uses simultaneously. This partially was a function of school board funding, the building floor plan, and the creative use of space by the JACS community. Building blocks of capacity can also be found in the diverse assemblage of not only investors of money, time, and talent, but also the freedom to contribute episodically, short-term, or long-term. All of the ingredients were aligned by the leadership to result in JACS's practically manifesting their community school vision.

## References

*The Adamite.* (April 1949). *Adams banking record 1948–1949.* Coatesville, PA: James Adams Community School.

Adams Community School. (April 29, 1955). *Annual spring operetta: A night in Spain.* Coatesville, PA.

Adams Community School. (February 5, 1954). *Twenty-eighth annual Evelyn Lawrence Oratorical Contest.* Coatesville, PA.

Adams Community School. (June 4, 1954). *Ninth annual induction ceremony: Adams chapter of the National Junior Honor Society.* Coatesville, PA.

Adams Community School. (June 5, 1951). *Graduation exercises.* Coatesville, PA.

Adams Community School. (June 7, 1949). *Graduation exercises.* Coatesville, PA.

Adams Community School. (June 8, 1948). *Graduation exercises.* Coatesville, PA.

Adams Community School. (June 13, 1956). *Move-Up Day.* Coatesville, PA.

Adams Community School. (May 25, 1956). *Achievement banquet.* Coatesville, PA.

Adams Gazette. (June 1952). *Objectives and Philosophy of the Adams school* (Vol. 27). Coatesville, PA: Adams Community School.

Anderson, T. J. (1950). Adams through the years. *Adams Gazette, 25*(2), 9.

Anderson, T. J. (December, 1953). *Coatesville City School district minutes.* Coatesville, PA: Chester County Historical Society Archives, Coatesville City School District.

Blank, M. J., & Berg, A. (2006). *All together now: Sharing responsibility for the whole child.* Washington, DC: Association for Supervision and Curriculum Development.

Blank, M. J., Melaville, A., & Shah, B. P. (2003). *Making the difference: Research and practice in community schools.* Washington, DC: Coalition for Community Schools, Institute for Educational Leadership.

Britton, J. O., & Britton, J. H. (1970). Schools serving the total family and community. *The Family Coordinator, 19*(4), 308–316.

Chester County Historical Society Archives. (April 4, 1955). *James Adams Community School: CIHU.* Coatesville, PA: Coatesville Record.

Chester County Historical Society Archives. (April 1954). *Our schools in action: James A. Adams school* (Vol. submitted by T. J. Anderson). Coatesville, PA: Coatesville City School District.

Chester County Historical Society Archives. (December 13, 1965). *Paul Palmer.* West Chester: Daily Local.

Chester County Historical Society Archives. (December 31, 1945). *A song recital: Roland Hayes.* Coatesville, PA: Coatesville Record.

Chester County Historical Society Archives. (December 1953). *Our schools in action: James A. Adams school* (Vol. submitted by T. J. Anderson). Coatesville, PA: Coatesville City School District.

Chester County Historical Society Archives. (December 1954). *Our schools in ACTION: James A. Adams school* (Vol. submitted by T. J. Anderson). Coatesville, PA: Coatesville City School District.

Chester County Historical Society Archives. (February 10, 1954). *William H. Ransom.* Coatesville, PA: Coatesville Record.

Chester County Historical Society Archives. (February 19, 1949). *Traveling toy exhibit.* Coatesville, PA: Coatesville Record.

Chester County Historical Society Archives. (February 19, 1952). *William H. Jones.* Coatesville, PA: Coatesville Record.

Chester County Historical Society Archives. (February 23, 1944). *Piano recital Philippa Duke Schuyler.* Coatesville, PA: Coatesville Record.

Chester County Historical Society Archives. (February 1954a). *Health in our schools: Health and physical education at Adams* (Vol. submitted by Gertrude Shoemaker). Coatesville, PA: Coatesville City School District.

Chester County Historical Society Archives. (February 1954b). *Our schools in action: James A. Adams school* (Vol. submitted by T. J. Anderson). Coatesville, PA: Coatesville City School District.

Chester County Historical Society Archives. (January 29, 1944). *Junior NAACP organized.* Coatesville, PA: Coatesville Record.

Chester County Historical Society Archives. (June 4, 1935). *James Adams town meeting.* Coatesville, PA: The Coatesville Record.

Chester County Historical Society Archives. (June 6, 1945). *World War memorial center.* West Chester: Daily Local.

Chester County Historical Society Archives. (June 6, 1956). *Leonard E. Dorsey: Who's who in American education.* Coatesville, PA: The Coatesville Record.

Chester County Historical Society Archives. (June 17, 1954). *Leonard Dorsey.* Coatesville, PA: Coatesville Record.

Chester County Historical Society Archives. (June 1952). *Objectives and philosophy of the Adams school.* Coatesville, PA: Adams Gazette.

Chester County Historical Society Archives. (March 6, 1954). *James Adams Community School: Winter carnival.* Coatesville, PA: Coatesville Record.

Chester County Historical Society Archives. (March 22, 1954). *James Adams Community School.* Coatesville, PA: Coatesville Record.

Chester County Historical Society Archives. (March 1954). *Our schools in action: James A. Adams school* (Vol. submitted by T. J. Anderson). Coatesville, PA: Coatesville City School District.

Chester County Historical Society Archives. (May 17, 1944a). *James Adams Community School.* Coatesville, PA: Coatesville Record.

Chester County Historical Society Archives. (May 17, 1944b). *James Boyd Memorial Library.* Coatesville, PA: Coatesville Record.

Chester County Historical Society Archives. (May 24, 1955). *Annual budget 1955–56.* Coatesville, PA: Coatesville City School District.

Chester County Historical Society Archives. (May 27, 1952). *James Adams field trip to Pomeroy Farm.* Coatesville, PA: Coatesville Record.

Chester County Historical Society Archives. (May 28, 1955). *Dr. Harvey E. Vanderslice.* Coatesville, PA: Coatesville Record.

Chester County Historical Society Archives. (May 1954). *Our schools in action: James A. Adams school* (Vol. submitted by T. J. Anderson). Coatesville, PA: Coatesville City School District.

Chester County Historical Society Archives. (November 1, 1946). *James Adams Community School sponsors school lunch program.* Coatesville, PA: Coatesville Record.

Chester County Historical Society Archives. (November 13, 1946). *James Adams Community School.* Coatesville, PA: Coatesville Record.

Chester County Historical Society Archives. (November 1953). *Our schools in action: James A. Adams school* (Vol. submitted by T. J. Anderson). Coatesville, PA: Coatesville City School District.

Chester County Historical Society Archives. (November 1954). *Our schools in action: James A. Adams school* (Vol. submitted by T. J. Anderson). Coatesville, PA: Coatesville City School District.

Chester County Historical Society Archives. (October 10, 1945). *James Adams Community School.* Coatesville, PA: Coatesville Record.

Chester County Historical Society Archives. (October 17, 1946). *Rev. James H. Brown Protests in Washington, D.C.* West Chester: Daily Local.

Chester County Historical Society Archives. (October 18, 1950). *James Adams Community School.* Coatesville, PA: Coatesville Record.

Chester County Historical Society Archives. (October 19, 1948). *James Adams Community School.* Coatesville, PA: Coatesville Record.

Chester County Historical Society Archives. (October 22, 1955). *Dr. Thomas D. Clayton.* Coatesville, PA: Coatesville Record.

Chester County Historical Society Archives. (October 30, 1946). *Rev. James H. Brown to Pastor in Erie County.* Oxford: Oxford Press.

Chester County Historical Society Archives. (October 1953). *Our schools in action: James A. Adams school* (Vol. submitted by T. J. Anderson). Coatesville, PA: Coatesville City School District.

Chester County Historical Society Archives. (October 1954). *Our schools in action: James A. Adams school* (Vol. submitted by T. J. Anderson). Coatesville, PA: Coatesville City School District.

Chester County Historical Society Archives. (September 23, 1950). *James Adams Community School.* Coatesville, PA: Coatesville Record.

The Children's Aid Society. (2001). *Building a community school* (3rd ed.). New York.

Clapp, E. R. (1971). *Community schools in action.* New York: Arno Press & The New York Times.

Cuff, B. (2007). *Interview.* Coatesville, PA: J.W. Richardson.

Davis, A. (2005). *Alumni move to have James Adams School bell put on display.* Retrieved September 6, 2008, from http://chestercountyreporter.com/100schoolbell.html.

Dryfoos, J. G. (1994). *Full-service schools: A revolution in health and social services for children, youth, and families.* New York: Jossey-Bass.

Dryfoos, J. G. (2003). A community school in action. *Reclaiming Children and Youth, 11*(4), 203–206.

Dryfoos, J. G., & Maguire, S. (2002). *Inside: Full-Service Community Schools.* Thousand Oaks, CA: Corwin Press.

Dryfoos, J. G., Quinn, J., & Barkin, C. (2005). *Community schools in action: Lessons from a decade of practice.* Oxford: Oxford University Press.

Harris, M. M., & Hoover, J. H. (2003). Overcoming adversity through community schools. *Reclaiming Children and Youth, 11*(4), 206–211.

Keith, N. Z. (1999). Whose community schools? New discourses, old patterns. *Theory into Practice, 38*(4), 225–234.

Kronick, R. F. (2005). *Full service community schools: Prevention of delinquency in students with mental illness and/or poverty.* Springfield: Charles C. Thomas.

Morris, J. E. (2008). Research, ideology, and the brown decision: Counternarratives to the historical and contemporary representation of Black schooling. *Teachers College Record, 110*(4), 713–732.

Robinson, B. G. (2005). *James Adams school alumni hold annual picnic.* Retrieved September 22, 2008, from http://www.zwire.com/site.coatesvilleledger.com/.

Sanders, M. G. (2001). The role of "Community" in comprehensive school, family, and community partnership programs. *The Elementary School Journal, 102*(1), 19–34.

# 6

# In Their Own Words: Reflections from Alumni and Faculty

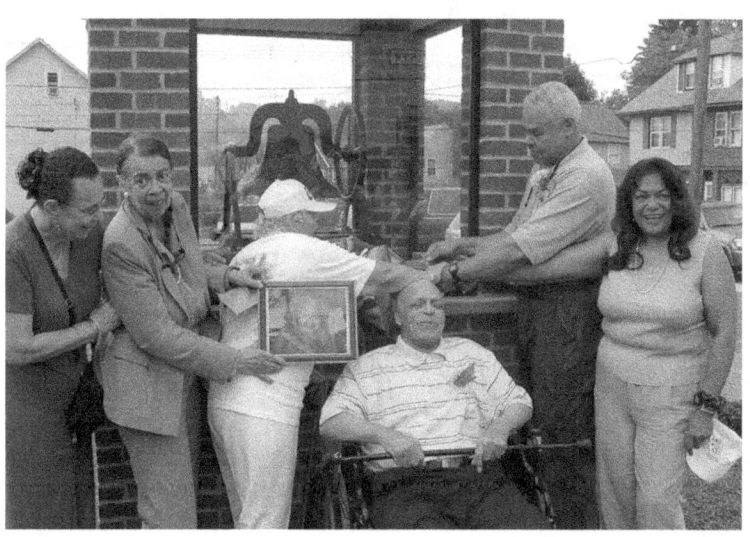

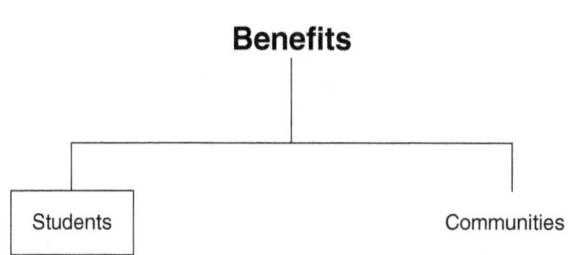

> I've always felt very blessed that I attended James Adams School.
> – (Gwendolyn Edwards, words from a survey)

> CIHU inspired me to want to do my best in and out of school. The philosophy was a constant reminder that even I could pursue my dreams no matter how far out of reach they may have been.
> – (Paulette Horne, words from a survey)

Academic deconstruction of full-service schools can run the risk of taking on an aura of the abstract. Similarly, accomplishments of students, administrators, teachers, and community members in this sort of exercise can seem unattainable and impossible to replicate. Despite these tendencies, it is important to remember that irrespective of intersecting risk factors, James Adams managed to leave its imprint on the hearts and lives of faculty, students, and its community. Paulette and Gwendolyn along with many others remind us of the potential of full-service community schools, particularly given the persistent hindrances to optimal achievement. In an effort to drag the abstract into the tangible and attainable, vignettes from the varied perspectives of the many James Adams community informants will be included here. Voices of the Andersons, students, and others are composites of the interviews, surveys, and documents that informed this study are set forth as vignettes capturing the essence of life in Coatesville and JACS. The chapter concludes comparing accounts of JACS's to the literature-based precepts of student and community benefits.

### Opening Bell at James Adams Community School

Imagine a school full of energy and activity palpable as you enter the front doors. The very imposing figure walking the halls ensuring students found their way to the proper places and disciplining when necessary was Mr. Anderson. His size alone encouraged compliance though once you knew him better one found a fierce advocate for educational and economic opportunity. Down the hall are classrooms where French, Latin, German, and Spanish are taught. Traditional content such as mathematics, English, and social studies are instructed in such a way that students did not feel purposefully subjected to a separate and supposedly unequal educational experience. Woven into every assignment and contest were opportunities for students to identify their talents, and build in themselves the confidence their instructors already believed they possessed.

Posters and announcements plaster the walls heralding upcoming events, accomplishments of students, and reinforcement of academic lessons. On any given evening there might be a mother–daughter tea, dance recital, oratorical contest, operetta, a play, famous visitors, or an award ceremony of some kind. As a student there was never a lack of things to do—rather the problem was prioritizing what one wanted to do most. Parents and elders of East End were not left out. There were almost as many activities for them to partake in as in youth. If for some reason you still didn't find anything to suit your fancy, you could either suggest a class or put your skills to good use supporting a planned event. Learning opportunities alone, though central to much of the programming, would have created an unbalanced schedule; so intergenerational leisure activities were plentiful. Movies, intramural sports, tutoring, and game nights were all designed to provide enjoyable outlets for students and adults to participate in together. Mixing in relaxed settings built a camaraderie that transcended age and professional affiliation.

Walking the halls during the day, one could sense the coexistence of love, discipline, and high expectations. Mr. Anderson was rarely in his office during the day; instead glimpses of him could be captured in doorways and the back of classrooms. In fact, "he was always in the halls directing people." His frame alone must have been enough to negate the lure of mischief. It was almost an oxymoron that someone so tall and broad would possess such a genteel manner. However, those who found themselves awaiting discipline learned that beneath that smile could be the sternest of principals. Discipline meted out either with the rumored paddle or stern warning was wrapped in a genuine concern for student well-being. Youngsters are experts at deciphering fraud and their hearts told them there was no mistaking Mr. Anderson's serious chastisements for compassion. So even in retrospect, students recall, "He was firm, friendly and a concerned person. He stressed that we were to do our best at all time," and, "We knew the rules and Mr. Anderson enforced he rules. He was tough but fair."

A peek into a classroom allayed any notions of substandard curricula. Yes, the books were older and in worse condition than those in the White schools. Yes, funding for the facility was not equitable, so the creature comforts did not parallel White schools. Offsetting all of that was the professional preparation in concert with the creativity and passion for teaching. Armed with undergraduate and graduate degrees from highly respected institutions teachers were empowered by Mr. Anderson to craft exchanges with students that did more than relay the rudiments of subject matter. Science, math, English and all the other courses were to be additive and relevant to the lived experiences of children. Eurocentric precepts were necessarily taught as part of a rounded individual's training. Yet this alone would have left students without views of themselves in the annals of history, art, and literature. So as they read Shakespeare they also read Zora Neale Hurston. Great scientists and philosophers such as Sir Isaac Newton and Aristotle were partnered with Drs. Charles Drew and Howard Thurman. Reinforced in the teaching–learning exchange was the expectation that students held the potential to achieve whatever they imagined if they were willing to work hard for it.

Moving from class to class the hum of activity and energy was mixed with anticipation of evening activities. Eavesdropping in the hallway as children moved from class to class one can imagine the plans being hatched. "What are you doing tonight?" "Are you kidding, I am going roller skating right after my homework is done. Last week I figured out how to turn around without falling down. That really impresses the girls." "Ha, you are so funny. I am going to shop because we are going to finish those chairs we started with Mr. Palmer last week. My mother can't wait to see it when we are finished."

Small groups of teachers huddle discussing the nominees for this year's safety patrol. "Barry has worked hard to improve his grades and would be an outstanding role model for the younger children." "I agree, what about Gwendolyn? She is another great candidate. Did you hear her sing last week during the National Junior Honor Society induction ceremony?" "Yes, she will represent us well in two years when she moves up to Scott [high school]. Speaking of Scott, did you hear that Lee Brown became the first Negro elected to student council? We will have to make sure that is included in next month's *Concern* and *The Adam Gazette*." The teachers look over the nominees, vote, and decide that the three open safety patrol positions will be announced in the morning. Scurrying back to their rooms, they collect papers and grade books and leave for home. There isn't much time to spare because most of them are due back at the school in a few hours. Some stop to visit a family or two on the way home, others go straight to the teacher's residence or home to their families to grab a bite to eat, and perhaps squeeze in a short nap.

At home mothers, fathers, and extended family members await the arrival of their children. They too are well aware that the time between school closure and community center activities beginning is short and precious. There is talk of how the day went and upcoming projects. Pomeroy Farms are the focus at one family's table as Jimmy Barksdale excitedly explains the impending field trip. "We get to walk around the farm and collect samples to bring back to class. I wonder if I will be able to catch a salamander or frog." "How did you do on your math test today?," his father asks. "It was fine, I think I got two wrong."

At the Andersons, just as in the other homes of East End, dinner was being prepared and their daughters setting the table for dinner. "Ila, make sure that you set an extra place at the table. Mrs. Moody is joining us today," Anita instructed. The girls easily accommodate another member to the family table as they are quite used to their home being as much of a community hub as the school. They didn't mind though; amazingly they never felt as if they were in competition with Coatesville for their parents' attention. They were quite secure in the knowledge that they were loved and in many ways precisely because they were Andersons, were the children of the community with "aunts" and "uncles" on every corner. Let's stop by 431 Cedar Street to get a sense of how plans for JACS take shape.

**Leading by Example**

It is late afternoon and the preparations for the busy week are in full swing. T. J. steps out for a meeting of the Coatesville Interracial Committee and before you see him you hear, " 'Nita...'Nita...." Anita answers, "I am helping Jean get ready for the oratorical contest." Though the winter carnival has consumed much of their time this week, there is always another program ahead. The following week is the twenty-eighth annual Evelyn Lawrence Oratorical Contest and the dance club's program, "The Age of Anxiety." Mrs. Anderson is quite pleased that so many people from the community and mostly faculty agreed to be patrons of the program and plans to use the proceeds for a scholarship.

Jean and Ila were not the only children of the Andersons. Their brother, T. J., Jr., a neophyte musician, who would one day become a world-renowned composer, was off in college and always very much a part of everyone's consciousness. Ila practices the piano and in the background one could hear Jean practicing Walt Whitman's "Oh Captain, My Captain":

> O Captain my Captain! our fearful trip is done,
> The ship has weathered every rack, the prize we sought is won,
> The port is near, the bells I hear, the people all exulting,
> While follow eyes the steady keel, the vessel grim and daring;
> But O heart! heart! heart!
> O the bleeding drops of red,
> Where on the deck my Captain lies,
> Fallen cold and dead

"No, no," Anita prompts, "you have to pause for effect after each comma and then crescendo to the 'heart! heart!'," Anita prompts. "Okay Momma," Jean dutifully replies. Anita muses over the selections for the contest that represent a true cross section of American literary genius. Works of authors such as Foss, Longfellow, and Whitman along with Paul Lawrence Dunbar and Langston Hughes were included in this year's contest. Anita smiles to herself knowing that companion recitations of African American and

Eurocentric classics planted yet another seed of confidence in students. She scurries into her room to put the final touches on lesson plans for the week before dinner.

"How are the plans for the carnival coming?," Andy (aka T. J.) asks as he rummages in the icebox (what refrigerators were called then). "What's for dinner?" "Just give me a minute and I will be right down," Anita responds as she switches gears like working mothers so often have to do. She stares at the exercises with a pen in hand pondering a more creative way to make Shakespeare relevant. T. J. walks into their bedroom only after he pokes his head in to see Ila and Jean chatting as sisters often do about this and that. "Hi Daddy," Jean says almost before she actually sees him. "We can't figure out this math problem." He pauses a moment after which he responds, "T-H-I-N-K girls. I know you can figure it out." As is his custom, he throws up his right hand before leaving, which by the way is large enough to palm a basketball, and reminds the girls, "Everything is alright." The truth is that Mr. Anderson was a man of few words and volumes of action, which is why his repetitive phrases carried such meaning. Spelling out T-H-I-N-K was one such example. Another was the seemingly simple gesture of throwing his hand in the air, smiling, and saying, "Everything is alright." When he said it, you believed everything really was all right.

"Andy have you talked with Mrs. Gibney to make sure she has everything she needs for the carnival? There are so many details we have to see to before Friday. Let's see, there is the moving of the furniture, food for the booths." "Anita, you know I handled that last week." The winter carnival is scheduled for Friday and it is a major fund-raiser. Without it much of what was accomplished would not have been possible. T. J. continues, "I need to finish editing this issue of *Concern* so it can go to press tomorrow. Jim Brown wrote a great article as usual. A new co-op store is planning to open in town and so he thought it would be a good idea to explain the history of co-ops. William Muthard [the principal of the integrated Scott High School] wants to talk about our strategy if the *Brown v. Board of Education* case is resolved in our favor. There will be teachers to reassign, as well as students and I agree with him we need to be in a position to recommend a transition plan we believe will be equitable. I had my doubts about him when we returned to Coatesville; he has proven to be sincere in his desire to support integration."

Anita turns her attention to fixing dinner, almost running into T. J. in the hallway, "I think you hit the mark and your article 'Segregation is Un-American' is quite good too I must say. My favorite section was when you cited the National Education Association's decision to oppose further spreading of segregation, which in theory still retains parallelism in schools where it currently exists."

Andy settles himself at the kitchen table, "I am thinking that a practical article about how to improve interracial relations will be of help to Mr. Muthard (principal of Scott High School and member of the Interracial Committee). During the last meeting it became apparent that many of our White friends are perplexed when they unintentionally offend Blacks so we thought we would make helpful suggestions and affirmations like congratulating theater and restaurant owners who do not segregate, and consider hiring Negroes in other than menial positions at a fare wage. We also suggested they conversely not frequent segregated establishments and suggested they be mindful of expressions and acts that Negroes find offensive. There is nothing like references to Southern mammies, rubbing a Negro's head for good luck, or using terms like spade or eight ball to offend someone. The point of course is not to be patronizing but rather informative. What did you think?"

"Good idea and very timely. I meant to tell you the Farcross family is doing much better. Mr. Farcross is still looking for work but I think he has a good chance of being hired at Lukens. The children found clothes that fit them nicely in the CIHU closet and food pantry gift was on time. We need to make sure young Paul gets some extra tutoring in French." With a hearty laugh that could be heard two rooms away, T. J. comments, "Paul Farcross isn't the only one that needs help with French. All the Hastings boys are interested in is track. Pop Ransom promised to make sure they finished their work before they set foot on the field."

"Come, let's take a break and eat dinner I think I hear Mrs. Moody knocking at the door." The family sits down at the table and as is their custom sings grace before eating, "God is great and God is good; and we thank Him for this food; By His hands must all be fed; give us Lord our daily bread; give us Lord our daily bread. . . . Amen". Conversation covers events local, national, and international. Attention soon turns to the dilemma facing their guest and as the adults strategize the daughters slip away from the table, clearing the dishes.

In no time the family piles in the car and is headed back to Adams. No sooner than they cross the threshold each scatters: Jean to the gym, Ila to the teen rap session, Anita to the winter carnival planning meeting, and T. J. to his usual mobile post, in and out of classrooms. It seems they had only arrived when it is time to go home and prepare for the next day.

When the children are in bed and the couple has time to themselves ideas take on practicality and plans are hatched. There were always more things to do than they personally had time for so part of the challenge was identifying talented and committed individuals who could aid in the work. T. J. and Anita donated more than their intellect, time, and money. Leading by example is their most enduring legacy. Finishing touches on the winter carnival are on their minds tonight. "How much do we need to make this year Andy?" "Well, a bit more than last year because we need to add a shop class for the girls and Mr. Palmer is going to need supplies. Also, Pop Ransom needs transportation money for the track team since we aren't running with Gordon any more." Anita piped in, "Mrs. Gibney is baking and you know not a single cake or pie she makes will be left over. In fact, we had better buy ours before the carnival starts." They both laugh as T. J. leans over to turn off the light.

### Teaching from the Heart

Moments before the alarm goes off, Betty awakens and smiles. "I still can't believe I am teaching in the school I attended as a child." She had recently returned to Coatesville from college and graduate training and Mr. Anderson had offered her a job teaching second-graders. Her parents were so proud she was a teacher, particularly since she comes from a line of educators, her father being the principal of the nearby Hayti Colored school for years. Now it was her turn to teach, but things in Coatesville had changed since she left. The town had always been at least partially integrated, because all of the town's teens attended tenth through twelfth grades together. Don't misunderstand, everyone knew the respectable boundaries between Blacks and Whites and rarely crossed them. Amazingly she had no problems worth remembering once she left the safe cocoon of James Adams, and she knows why: she had been equipped to handle the rigors of college preparatory coursework and her teachers regularly checked on her progress. Now in her adulthood a push for integration nationally seems to be on everyone's mind. Whispers of an imminent Supreme Court decision and the potential consequences weighed heavily on the minds of Negros and Whites alike.

She decided her energies were best spent tending to children and making sure that they were well equipped for life outside their little section of Coatesville. If this goal was to be met it would take more than traditional school hours, administrative support, and innovative pedagogy. It would take everyone working together, which was part of the appeal of the Andersons' return. Their reputation for getting things done preceded their second stint as heads of the school.

As she moved albeit slowly to the kitchen for that all important first cup of coffee, pictures of the Farcross family floated through her mind. They were like so many families sequestered in the East End: dad trying to get work, mom picking up odd jobs as she could, and the three children. My they were bright, she thought, as she took her first sip. Little Yvonne was in her second grade class. Betty gathered her materials for the day and walked the familiar path to school, acknowledging others on their way to work. Many of the fathers worked either at the steel mill or the VA hospital. While Coatesville had always been her home, she wondered how long young professionals like herself would keep coming back after they were educated if opportunities for jobs were so limited. She was one of the lucky ones, because to work at JACS was to have respect in the community, earn a livable wage (almost the same as White teachers doing the same work) and making a difference in the lives of children. She just couldn't imagine being anywhere else or doing anything else.

Walking through her "to do" list she thought, "Let's see.... I already assigned a student to be in charge of collecting for the school bank. Barry Farcross earned extra merit points for his CIHU project of reading to children in her class and she must remember to write that down. Her students were going to sing 'American the Beautiful' at the oratorical contest so they needed to practice; oh and she needed to remind her students to bring items to donate to the clothes closet and food pantry. It gives even her second-graders such a sense of accomplishment to know that despite their age, they can help the community too."

Reflecting back on those days, Betty considers the disposition of contemporary teachers a mystery. "In my day when you decided to teach you knew that was a pretty consuming job, but more than a job it was your joy to help children. We thought of home visitation as part of the job. I learned important things during those visits [they] helped me target assistance for my students. It's funny now to hear folks talking about how important a child's home life and health are to their schooling. Mercy, we knew that more than 60 years ago. That's why we didn't mind going to students' homes or going back to school. We saw the impact of building relationships with families and sharing work and recreational time with them. So many benefits accrued to our students."

"As a matter of fact, once they were our students, they were always our students. When they moved up to Scott (High School), we would check on them to make sure they were in the right classes. Not everyone was fair minded so we needed extra sets of eyes to keep White teachers and administrators dealing honestly with our alumni. They were our students even after they graduated from Scott and went to college. It wasn't uncommon for us to work to find scholarships to support college costs. In fact, we actually started building their scholarship base at James Adams even though at the time they had only completed ninth grade. Scholarships at that juncture made clear our confidence in our children's intellectual potential."

"I am convinced that our students did well academically because we expected them to. Sure, everyone has different abilities and interests, but our job as teachers and members of the same community was to make sure they were pushed intellectually to achieve as much as their ambition would allow. I think there are so many reasons JACS was the vibrant center it was, metaphorically it was as though all the stars lined up to make it so. First we had the Andersons, who were very well respected in Coatesville, by Negroes and Whites. They were a bit like royalty mostly because they walked the walk. Do you know what I mean by that? If we needed donations for food or clothes for the clothes closet or pantry, the Andersons were the first ones to give. If we needed money to run a program Mr. Anderson could get businesses around town to donate money and resources to us. It was pretty impressive. We all did our part too. We [the faculty and staff] pooled our resources and gave scholarships, were patrons of programs like the Operetta and helped with the winter carnival."

"When the library burned and [fire] destroyed all the books and a few classrooms, all the classes and teachers were scattered across town until the repairs could be made. We mostly taught our classes at the end of the white schools' day. That was a trying period, borrowing space and supplies from whites; I could hardly wait until we were back in our own school again. We worked together by talking to friends and family members about donations and by the time we reopened the library we had over 2000 books. There were so many books and the investment in restocking came from such a cross section of the community that when it opened it was called the James Adams Community Library. Everyone could borrow books, not just students."

Mrs. Cuff-Stokes doesn't understand teachers nowadays, who are at times portrayed as professionals who want to clock in and out of work. She has no frame of reference for it. Teaching in her time was like a calling something deep inside that compelled you to serve. Given that frame of mind, visiting the homes of students and getting to know parents better on their turf wasn't a burden—it was just part of the job. Staying after school to help with clubs, classes, and tutoring was also what she knew she was signing up for when she returned from college.

The other reason she speculates that teachers didn't mind working so hard was the freedom they were afforded to develop innovative pedagogies, to include opportunities to create interdisciplinary projects. Whenever you had an idea, you were not greeted with why it couldn't be done, but rather the question, how do you think we can make this happen? Also, there was a sense among the teachers that their work was appreciated and respected, not just by parents, but also the administrators and other teachers. They each had a part in co-creating JACS and because of that had more of a vested interest in its successful preparation of students for a changing world.

The best thing about this week was the winter carnival on Friday, Betty mused. "What fun. Oh, that reminds me, I need to get the children ready for their class booth...."

## A Child Will Lead them

Paul is running because as usual he was almost late to school. Mr. Anderson will be standing at the door as always and the last thing he needs is another tardy and the inevitable meeting of his seat of knowledge with the end of Mr. Anderson's paddle. Static ekes out of the public address system "Good Morning. Today is Tuesday, February 17th, 1954. Everyone rise to recite the pledge allegiance 'I pledge Allegiance to the Flag, of the United States of America, and to the republic for which stands, one nation under God Indivisible with liberty and justice for all'. Now Jean Anderson will lead us in singing our CIHU song...." Everyone joined in, some with more vocal prowess than others. No one seemed to mind for example that Lee sang completely off key. "You may be seated." Paul pulled out his French notes to review for the quiz today as the rest of the announcements were read. "The health and dental clinic are open today. All second-graders will report to Dr. Clayton's office immediately after the announcements. Tonight Pop Ransom and the Adams basketball team will play Downingtown. Be sure to come out and give them your support. In other news, Friday is the annual winter carnival. All class displays, and booth preparations must be completed by Thursday. Lastly, if [you] are not going home for lunch today, the cafeteria is serving fried fish and macaroni and cheese."

Paul paused from rehearsing his French vocabulary as he recalled his latest act of CIHU. On his way home from school one day last week he saw Mrs. Prout struggling to carry her groceries. He was in a hurry, but thought about how much she reminded him of his grandmother. He helped her all the way to her house and off he went. He had forgotten about it until Mrs. Keith told him that Mrs. Prout sought her out in church and as a result points were added to his merit score. The way he was racking up points he might be in contention for the big award at the end of the school year.

He disciplined his wandering young mind to focus on the French vocabulary. He loved the grin that would predictably creep across his father's face when he brought home good grades. Besides, if he didn't keep the grades up, Pop Ransom would "motivate" him by making him sit on the bench the whole game. That was enough to keep most of the fellows studying. The bell rang and more than 500 students began their class work. After the quiz (he thought he did fairly well on), he moved through the rest of his classes—mathematics, science, English, and then home for lunch. Once he returned, there was history, Spanish, physical education, and woodshop class. He particularly liked shop because students could make furniture to sell or give to those in need. This week they were working on new bookshelves for the library. Even after the library reopened after the fire, books kept pouring in. Mr. Palmer had confidence that the students could make shelving every bit as good as commercial stores and that made them proud to help.

Paul gathered his brother and sister (Donald and Sis) and went home at the end of the school day. As it turned out on most nights they weren't home long after dinner, because there was always something going on at the school. His parents didn't mind though, because they knew they were well taken care of by the adults at JACS. Drifting off in his imagination during dinner, he started thinking that he wanted to be lawyer like Thurgood Marshal. Imagine standing up for your people in front of all those influential officials. Mr. Dorsey told them in class today that Mr. Marshal earned his law degree at Howard University. That was the same school the Andersons and a couple of his other teachers attended. That must be some place! He decided at that moment, "I am going to Howard University and I am going to make my parents and my teachers so proud. Who knows, I may even become a judge one day."

### CIHU Spills Over

Not everyone that benefited from JACS's outreach was African American. There were members of the White community who basked in the CIHU spillover. Carmin knew the Andersons and the James Adams School from this vantage point. He and his family attended the Seventh Day Adventist Church on the edge of town where T. J. was a member. Mind you, they were the only White family in the congregation, but this didn't seem to trouble Carmin's parents at all. A grin still creeps cross Carmin's face when he says, "there was only one way my father referred to Colored men"; he pauses and then says, "brother," and chuckles.

From his vantage point, though Coatesville was a segregated town, it was not hostile. He recounts the names boys called each other. The Blacks referred to Whites as "crackers," because, he presumed, they were white like saltines, and his friends referred to Blacks as "graham crackers" because they were brown. More vocabulary was clearly available, and surely there were exceptions on extreme ends of the acceptance continuum. Yet, based on his experience, Negroes and Whites coexisted amicably.

And enjoying each other's company as social and chance gatherings would allow.

Carmin's perspective might seem inappropriate as we reconstruct life at James Adams; however, his story provides additional evidence that the Andersons led by example, lived CIHU, and made judgments about people based on character as opposed to race or ethnicity. Each such demonstration added to Mr. Anderson's gravitas, prompting investments in JACS that other Negro schools did not seem to enjoy. The White Seventh Day Adventist Church was a distance away and the Garnett family's means of transportation were not always reliable. The nearest Seventh Day Adventist Church was part of the Allegheny Conference and was a Black congregation. Though his family were the only White members to his knowledge, as the congregation grew so did the numbers of White families that would visit.

"T. J.," Carmin recalls, "was not just big, he enveloped a room. Most memorable was how warm, encouraging, and accepting he was in all his dealings. Though any of the more established congregations would have welcomed the James Adams principal as a distinguished congregant, he chose to attend a neophyte congregation held in a converted two-car garage. It amazed me how comfortable T. J. was in the unimpressive edifice; in fact it was impressive how comfortable he was anywhere with any company."

When asked about CIHU Carmin recounted that it was not a phrase articulated much external to the school, but telltale actions were evident. "T. J. was quick to aid anyone in need and was instrumental in sparking my interest in attending a well-respected preparatory school in Philadelphia and pursuing the ministry. Ever the team, Mrs. Anderson though a devout Methodist did not join the church during T.J.'s lifetime, but she often played the piano for services as only she could."

"Did you know T. J. would sometimes stammer? He only did it when he was excited. I can see him now, leaning forward in his chair teaching a Bible lesson in Sabbath school. It was inspiring," Carmin reminisces. Mr. Garnett's family would often ride home from church in the Andersons' car and by virtue of church affiliation came to know each other well. T. J.'s leadership in the church was not limited to children's lessons. When it became clear that the congregation needed a larger place to meet, he spearheaded their relocation. He found out that a nearby Lutheran assembly was selling their church, compiled information from the bank about the necessary down payment and mortgage terms, and shared the information with the congregation. When as members they doubted the odds of their being successful in such a large undertaking, it was T. J.'s confidence and encouragement that aided in the decision to purchase the property.

Acts of kindness irrespective of race were an Anderson calling card and contributed greatly to their ability to secure support for JACS from diverse quadrants of Coatesville society. Their commitment to bettering not only East End but the whole of Coatesville turned out to be a support magnet between the White- and Black-owned establishment and the school.

Carmin's father worked at Lukens and in one of the rides home from church Mr. Anderson mentioned Mr. Farcross and how much he needed a job. T. J. vouching for anyone's reliability and skill was as good a reference as one could have. Mr. Garnett, trusting T. J.'s judgment, committed to speak to the union representative on Farcross's behalf.

**Parental Pride**

Mr. and Mrs. Farcross don't know what they would have done without the JACS family. They were proud people and had moved to Coatesville from North Carolina in search of work. At home there had been much talk about the jobs available in the steel mill that paid well. There weren't any jobs in North Carolina that would help the couple provide a better future for their children, so they decided to leave. First, Mr. Farcross moved to investigate Coatesville and got temporary work at the mill. Housing for a man alone was a challenge to find, but soon someone told him about a boarding house for Negroes in the East End. He worked odd jobs, as well as any hours he could pick up at Lukens. A few times he worked side by side with Mr. Anderson to plant a huge garden in the back of 431 Cedar Street (the Anderson home). He sent as much money home as he could until his wife Rowena said there was no point in them trying to survive in two places. She convinced him that somehow they would manage and soon the couple found a small house to rent on Elm Street.

Elm Street, like other blocks in the East End, was composed of row houses. The narrow structures reached backward halfway to the next street, sometimes with small backyards. There was a front porch, a "front room" (living room), and a foyer where guests would be welcomed. Beyond the front room was a dining area leading back to the kitchen and back door. Upstairs were bedrooms and usually one bathroom. Basements could be converted to provide extra living space as well as attic crawl spaces. Given the era, the rowhouse provided ample space for a family of five. Each structure shared at least one common wall with the neighboring house.

Grateful is the best word to describe the way the Farcrosses felt about having found a place where they could finally be together again. That said life was not as easy as they had been led to believe. They made do as best they could, pooling the money from odd jobs to pay the bills. It was great news when the school's free and reduced school lunch program was announced. Unfortunately for Fred and Rowena, the news was met with only a partial sigh of relief. With three children, even the reduced amount needed for lunches was difficult to come by. Then the real godsend—the PTA heard about them and paid the difference between the government subsidy and the family contribution. Rowena was not ashamed to admit she had tears in her eyes that day. Then there was the matter of keeping growing youngsters clothed and the CIHU closet where they could find what they needed.

Perhaps the best thing about James Adams and their willingness to aid people in the community was how they helped. Rowena and Fred never felt patronized by faculty or staff. Food and clothing were available discretely and maintained a family's sense of dignity. One just slipped into the holding areas, picked out what one needed, and took it home. With each visit Rowena determined that one day her family would be the ones donating, but for now it was such a relief to know the children would not be hungry. In the meantime, there were lots of ways to help the school, which made it possible to give back to the institution providing so much tangible assistance. News that Rowena was a gifted seamstress didn't take long to reach the Andersons and before she knew it she was leading the "Make-Do and Mend" club. Sessions were a cross between tangible skill building in terms of sewing and cathartic bonding with women in the community.

Her husband eventually found a full-time job at Lukens, making her dream of giving something back to the institution that had helped her family so much a reality. He wasn't getting paid the same as the White men he worked beside, but compared to many other families they were blessed. He made enough money for them to buy a little house in Carver's Court and for her to stay at home with the children.

When she was a little girl she had dreamt of becoming a teacher, but her parents didn't have enough money to send her to college. At JACS she could do the next best thing. Extra sets of hands were always welcome at the school and she saw no point in sitting at home; she signed on as a homeroom mother and in no time she oversaw homeroom moms from all the grades. Homeroom mothers were in every kindergarten and lower elementary classroom. Their job was to assist the teacher with things like hygiene checks, collections for the James Adams Bank, and any additional support the teacher might need during the class day. By being in the classroom she saw first-hand the high expectations the teachers had for students and the ways class assignments were reinforced in after-school activities. No wonder her oldest son Paul had begun talking about going to Howard University and becoming a lawyer. As for her, she started to think it might not be too late for her to go to school. With the help of a few of the teachers, she submitted her application to Cheyney State Teachers College and she was waiting to hear the decision. In the meantime there was always more to do at JACS, like getting ready for the winter carnival at the end of the week. She turned her attention to embroidering James Adams onto the jackets that were going to be given away in the raffle.

## Beyond the Years of JACS

As an educational researcher, I have spent years advocating for more holistic interventions on behalf of school-aged children. Most of my work has linked social determinants of health to the learning readiness of children to urge public health and educational professionals to collaborate. Over the last few years I have been most intrigued by contemporary calls for schools to use their facilities more efficiently by making them available to the communities they serve. Public discourse about school choice, resegregation of neighborhoods, and calls for community engagement were themes in my work at the time I became aware that former students of my grandparents held annual reunions in the town of my youth, Coatesville.

There are times when the research calls you. This is not meant completely in a mystical sense, but rather in an acknowledgment that much research is prompted by a hunch, that intangible something that pulls our attention from other pursuits. My initial nudge was the frequency with which I noticed contemporary calls for innovative school collaborations, that is, full-service community schools. This was striking to me because though never a student at James Adams, I was well aware that my grandparents with few resources had managed to create this type of school over 50 years ago.

I should begin this researcher disclosure from the very start itself. My grandfather and grandmother always loomed larger than life to me. They were and continue to be two of the most remarkable people I have ever known. Though their physical departure robbed me of their presence, their contributions to our family and their educational and leadership legacy transcend decades. It is funny how the things that frustrate you as a child can bring you such comfort in adulthood. Growing up in Coatesville, my brother David and I never got away with any youthful prank because everyone knew our grandparents and parents. Before we could get home with an explanation, grandmother and my mother would be waiting at the door. Our two younger siblings though born in Coatesville, never experienced the full Coatesville experience as we did by virtue of their ages. Looking back, I am sorry they don't have as many memories as David and I amassed.

As an adult, I walked many times down the unkempt streets of Coatesville and voices of elders called out to me, "You look familiar... aren't you an Anderson?" I always smile because I am known and connected to this place.

I am a fourth-generation educator and I cannot help speculating that the nobility of this profession was instilled not only from what I heard, but what I witnessed. My earliest memory of T. J. Anderson, or Andy as my grandmother Anita called him, is of a man as gentle as he was imposing. His height and broad shoulders equipped him well for his athletic youth and were also perfectly suited for granddaughters to ride on. It was his and grandmother's demonstrated commitment to education that heavily influenced not only their children (all three of whom are educators), but also five of their ten grandchildren. I vividly remember conversations about how opportunities to acquire education conferred upon us the responsibility to help others in our community who had not had been similarly blessed. Yet, it was not so much the words admonishing us to pursue academic excellence and service as their example of how it could and should be done. Our Cedar Street home was perpetually open to all sorts of people, seeking the counsel or aid of my grandparents.

My memories of Coatesville rarely conjure images of segregation, though admittedly in the early 1960s formal segregation had been legally prohibited but vestiges remained. We still lived in fairly clustered racial neighborhoods slowly transitioning because change takes time. Our home on 431 Cedar Street was a huge edifice on a corner of an integrated block. There were professionals and blue-collar workers on the street, as I recall Dr. Howard (the town's Black dentist), several teachers, and fathers (both Black and White) who worked at the Lukens Steel Mill. By the time I started kindergarten, James Adams had been closed several years and Grandpa had retired. Grandmother, however, was the first Black woman to teach English in the integrated Scott High School. Though desegregation had begun, enough remnants of the cohesive Black community still existed, affording me some of the benefits my mother experienced of being one of the children of the community.

In some ways it is hard to go back to Coatesville now. It has suffered the fate of many former industrially based towns. Downtown, which used to be vibrant with stores of every type, resembles a ghost town. A couple blocks off Lincoln Highway (also known as Route 30) our church still stands (Hutchinson Memorial) and a plaque designating Atkinson Hospital (where I was treated for many a malady), a historic landmark. Fortunately, despite what my eyes observed, I rediscovered the Coatesville of my youth talking with Adams alumni and reading about people I knew growing up. Researching clue after clue helped me reconstruct the masterful creation of the community dimensions of James Adams.

With every antiquated newspaper articles, school board, and city council minutes, I inched closer to the true treasure trove of JACS programs, *Adams Gazette*s, yearbooks, and hand-written notes scribed by my Grandmother and donated to the Chester County Historical Society when James Adams closed its doors. It was as if T. J. and Anita knew that what they had accomplished would provide insight to future generations of educators.

Most relevant to this evaluation of JACS are the strategies that later were substantiated as "best practices" in the scholarly canon. As the state of segregated public schools in the nation is considered, it is remarkable how much has remained disturbingly similar to the pre-*Brown* era. The good news is that we can glean not just what the Andersons accomplished, but also how they managed to be successful as we piece together the role of leadership, partnerships, and organizational development.

All of that aside, the real legacy of JACS is embedded in the lives of students who emanate the deposits of teachers, staff, and the community dating back to their youth. Professional affiliations vary widely, with as one might predict a large contingent of teachers. As accomplished as some alumni are, gatherings are not marked by such things. What took root in grade school has survived: a sense of responsibility to contribute positively to whatever community one belongs. Even now the alumni award scholarships to Coatesville High School graduates going on to college. They worked together to create a historic memorial to Adams, through the encased school bell standing near the James Adams site. In keeping with tradition, some of the monument's craftsmanship is the handiwork of local vocational school students.

Capturing the real spirit of JACS, however, takes more than reading archived data. One needs to be in a room of Adams' alumni and hear them singing the CIHU song these many years later, "I'm calling CIHU, I'm calling CIHU, Why don't you answer me?". Just as their teachers and community answered, they have answered, and so have I.

### Highly Effective Community School Model—Student and Community Benefits

Previous chapters were devoted to the leadership, partnership, and organizational capacity dimensions of the Highly Effective Community School (HECS) model. Not so easily measured are the benefits of a highly effective school on students and the community. Markers such as community pride, commitment to lifelong learning, public service, and the influence of the school ethos years after students matriculated are not easily assessed. Proxies for these measures are community contributions that were woven into the school curriculum, educational attainment and career

**Table 6.1** Benefits of a Highly Functioning Community School for Students and Communities

| Student Benefits | Community Benefits |
|---|---|
| Combats stereotype threat | Increased youth connection to the community |
| Creates supportive ecologies around youth | Opportunities for intergenerational bonding and learning |
| Acts as a protective factor insulating against risk and promoting resilience | School is the site for community problem solving |
| Site for needed services at a location that improves access | Providing parents with academic support for their children |
| Extends learning opportunities | More effective use of the facility and publicly allocated resources |
| Increases parental involvement | |
| Increases child safety because of supervised activities | More effective schools are attractive to businesses and realtors |
| Bonding with caring adults | Preparing a better-educated workforce |
| Chronic and acute diseases are managed earlier and more efficiently | Providing adults social, educational, and psychological support |

paths of informants, the annual reunion event, and the scholarships that are awarded annually to Coatesville students in the name of JACS. Recapping much of what was discussed at length in Chapters 1 and 2, Table 6.1 summarizes the benefits that can accrue to students and the community when a community school functions optimally.

The vignettes earlier in this chapter provide vivid examples of how engaging the community can create positive spillovers for educators, students, families, and the broader community. Individualism and the notion that one (or one's family) is solely responsible for the measure of success denies the benefits of a communitarian framework. As per Dewey, the more effective communitarian approach to schooling presumes each child is important and worthy of investment. Stated another way, community engagement is paramount to the nurturing of well-adjusted youth and the benefits of community schools accrue to all members of a civil society (Dewey, 2001–1915). Conversely, children, who are not recipients of adequate human welfare intervention (health care, food, safety, housing security, and emotional support), are not positioned to perform well academically. Mentorship, enrichment programming, rigorous academic curricula, and opportunities to build relationships with caring adults are as critical to the teaching–learning exchange today as they ever were (Blank & Berg, 2006; Dewey, 2001–1915;

Freudenberg & Ruglis, 2007; Kana'Iaupuni, Donato, Thomson-Colon, & Stainback, 2005; Swanson, 2008).

Positive student outcomes (benefits) such as academic performance are possible in comprehensive schools despite the presence of risk factors. Evidence can be found in the school board minutes noting, the numbers of students graduating from JACS to Scott increased substantially after T. J. and Anita's return (Chester County Historical Society Archives, March 22, 1954). Attendance rates also improved during this period, improving the likelihood that more time in class would reap positive results. The establishment of the district's first Junior National Honor Society speaks for the academic standards in place at the school. Interviews further confirmed that the transition from JACS to Scott might have been socially awkward, as a function of moving from a segregated to integrated school; however, students felt well equipped to master the academic curriculum in the high school because of their rigorous preparation at JACS and the counternarrative deeply imbedded in the curriculum and after-school activities, which proved protective as students moved to new environments.

Table 6.1 provides a list of both traditional and nontraditional community school benefits for youth. Harkening back to both our discussion of schools as protective agents and the scenarios in this chapter, we find additional proof that full-service schools work. They can yield the outcomes we profess we desire, youth equipped to maximize their potential and committed to responsible civic engagement.

Evidence of community benefits can be found in the reduced delinquency rates of students, and the attendance at JACS events. It was not uncommon for 300–400 persons to be in attendance at events such as the winter carnival, graduation ceremonies, and evening classes. What is also clear is that JACS was considered a vital contributor to community well-being and vice versa. Without one the other would have suffered. This symbiosis proved additive to students as they were the ones who were the primary beneficiaries, as the micro- and mesosystems of their world conspired to create supportive developmental ecologies. Secondary beneficiaries were parents and other adults who received aid, not only as it directly pertained to their children but also as it met their needs. Social and morale-building activities clearly bolstered the health of individual families and their ability to seek and offer support to others. Each learning activity and accomplishment created another brick in the façade of James Adams until in the end a highly effective community school was born.

## References

Anderson, A. T. (1982). *Interview*. Coatesville, PA: David Turpeau.
Anderson, A. T. (1989). *Interview*. Christiana, PA: Angella T. Hayes.
Blank, M. J., & Berg, A. (2006). *All together now: Sharing responsibility for the whole child*. Washington, DC: Association for Supervision and Curriculum Development.
Chester County Historical Society Archives. (March 22, 1954). *James Adams Community School*. Coatesville, PA: Coatesville Record.
Cuff-Stokes, B. (2007). *Interview*. Coatesville, PA: J. W. Richardson.
Dewey, J. (2001/1915). *The school and society & the child and the curriculum*. Mineola, NY: Dover Publications.
Freudenberg, N., & Ruglis, J. (2007). Reframing school dropout as a public health issue. *Preventing Chronic Disease, 4*(4), 1–11.
Garnett, C. V. (2008). *Interview*. Washington, DC: J. W. Richardson.
Gibney, B. (2007). *Interview*. Coatesville, PA: J. W. Richardson.
Hayes, A. (2008). *Interview*. Washington, DC: J. W. Richardson.
Johnson, P. (2006). *Interview*. Coatesville, PA: J. W. Richardson.
Kana'Iaupuni, S. M., Donato, K. M., Thomson-Colon, T., & Stainback, M. (2005). Counting on kin: Social networks, social support, and child health status. *Social Forces, 83*(3), 1137–1164.
Kelly, L. C. (2008). *Interview*. Oakland, CA: J. W. Richardson.
Swanson, C. B. (2008). *Cities in crisis: A special analytic report on high school graduation*. Bethesda, MD: America's Promise Alliance and the Bill & Melina Gates Foundation.
White, J. A., & Prince, I. A. (2007). *Interview*. Silver Spring, MD: J. W. Richardson.

# 7

# Looking Backward to Plan Ahead: Lessons for the Community and Full-Service School Movement

A community school is established to perform a job of total education.... An integrated day and night program caters to all elements of the population and seeks to bridge cultural gaps which are largely the result of unfulfilled group needs. The ultimate objective of the programming is to develop good American citizens who will make positive contributions to society.

– *T. J. Anderson, Sr. (Anderson, November, 1953)*

Imagine that society has told you directly and indirectly that you are a second-class citizen, undeserving of equitable treatment and investment. Imagine seeing all around you evidence of disparities that nag you as you try to understand why they exist. In the absence of a counter-narrative that affirms your self-concept and prepares you in practical ways for the future, you are doomed to live a life below your potential. JACS was a haven that did not insult the intelligence of children by denying the potential impact of their needs, what they saw day to day, or its potential influence on learning readiness and achievement. Faculty and staff met students where they were and created a "world" where they could learn, experiment, and identify their giftedness in the safety of a supportive community.

Admittedly there are dilemmas that are unique to specific moments in time that preclude whole-scale replication of James Adam's strategies and programming. However, practices common in the twenty-first century that ignore all but the teaching – learning exchange are not yielding acceptable academic outcomes for many of America's children. Fractured communities and families; demands to be technologically wired and proficient; growing numbers of children for whom English is a secondary language; and a compartmentalization of professional and personal lives all are contemporary challenges. On the other hand, some of the difficulties that educators faced in African American communities during the time of segregation still prevail: under-resourced neighborhoods, poorly funded schools, parents with low educational attainment, poverty, and segregation (now *de facto*).

Similarities between the 1950s and 2009 are unfortunately greater in number than differences and the national economic stakes are arguably higher. Though for almost all racial and ethnic groups educational attainment and achievement have improved over the last half century, certain subsections of the population still have not matched or surpassed the attainment of White or Asian children. A lack of success in early grades has repeatedly been identified as a predictor not only of risk in K – 12 academic outcomes, but also of dropping out of school prior to completion (America's Promise Alliance, 2008). However it has also been well documented that despite the plethora of risks a child might experience, if there is a place where resilience can be fostered, grade-level achievement and school completion is attainable. One antidote for risk has been and continues to be full-service community schools.

As complex as the aforementioned challenges are, I propose altering direction requires not only governmental intervention, but also individual community alliances with schools committed to nurturing resilience in youngsters. Issues associated with maximizing educational potential and access to quality education are relatively speaking as rampant today as in the late 1940s and 1950s when JACS was open. Yet, in terms of prevalent practice in pre-*Brown* educational settings, the challenges to learning noted above were not acceptable excuses for poor academic performance (Hilliard III, 2003). The same holds true with effective teachers in the twenty-first century. It could be argued that the persistent "achievement gap" between Blacks and Whites is more a function of what Asa Hilliard (2003) referred to as an "opportunity to learn gap." Research inertia, that is, continually deconstructing student deficiencies, contributes to the gap by failing to elevate the inequities educators can render harmless with culturally appropriate interventions.

A danger of yet another manuscript espousing the impact of student deficits is the creation of pessimism about innate of potential and the perception that teachers and schools are powerless to make a difference in the lives of their students. A "one-size-fits-all" education model has perpetuated in some cases, and exacerbated in others, the stubborn "opportunity gap." We already know what works for children because there have always been "gap closers." As Ronald Edmonds notes, "We can, whenever and wherever we wish, teach successfully all children whose education is of interest to us" (Hilliard III, 2003, p. 163). Chasms between African American and Caucasian academic outcomes are deeply rooted in the historical incremental access to quality educational environments.

Teachers and schools despite dire predictions can facilitate and have facilitated high academic achievement in their pupils as exemplified in this evaluation. These professionals should not be considered enigmas, but rather architects with plans that should be evaluated and replicated. Rather than focusing on what students did not have, they concentrated on the ways their teaching could change the lives of youngsters. These educators were not dreamers, but rather research substantiates the efficacy of their philosophy. High achievement results quickly when professionals shift their energies from deficits to supporting and expecting high academic outcomes (Hilliard III, 2003).

Change, as well as insulating children from stereotype threats in African American communities have historically been instigated at the community level (Morris, 2004). This is also what makes lessons learned from JACS relevant in a contemporary context. JACS, like others of the segregation era, managed to thrive despite the limitations inequitable support might predict. School leaders would do well to extract not only what the Andersons, faculty, and the Coatesville community accomplished, but how they did what they did. We can start by demystifying historical success stories because to keep work in that space mythologizes what can be accomplished in our time. Stated another way we need to frame the work of the Andersons and their partners as, "Ordinary teachers and ordinary principals with extraordinary commitment and energies can transform ordinary schools, and even failing schools into islands of hope in a sea of despair" (Hilliard III, 2003, p. 158).

As we consider the applicable strategies of JACS to the twenty-first century, let us begin by revisiting the theoretical suppositions that support the community and full-service school model as better for many children than traditional schools. Urie Bronfenbrenner's ecology of human development (EHD) model reminds us that childhood development is heavily influenced by concentric systemic spheres imposing themselves simultaneously. For children to arrive safely to adulthood, or be well equipped to become contributing adults, the spheres more directly engaged with youth (microsystems, and mesosystems) need to support their optimal development. In the absence of caring one-on-one relationships with adults (microsystems) and supportive interactions between and among schools, homes, and the community (macrosystems), "dis-ease" with their environments threaten both tangible and intangible academic outcomes. Though educators do not control all dimensions of their students' lives, via the community school model they can serve as protective agents against disruptive ecologies through the creation of supportive service networks.

One of the first ways to circumvent failure or underperformance can be found in the caring adult relationships, so critical to the internal narratives that influence students' choices. We found for example in the contemporary data the ways stereotypic threats imposed by the macro-, exo-, meso-, or microsystemic levels can become self-fulfilling prophecies. Youngsters are astute synthesizers of both negative and positive messages relative to their predicted place in society. Messages that they are threats to civility or that they will underperform academically are internalized in the absence of counter-narratives predicting success. As educators we have little control over the macro- and exo-systems that promulgate negative outcomes for our students. We do, however, control dimensions of the meso- and microsystems that can build self-esteem, self-confidence, and personal responsibility.

Counter-narratives combating stereotypic threats and disruptive ecologies (the combined risk factors to which youth are exposed) in the case of JACS were powerful in redefining possibilities. Students were not patronized with verbiage alone. Instead, high academic standards combined with the holistic web of support services, activities where students could succeed (sports, fine arts, crafts), and regular acknowledgment of their individual accomplishments (such as the merit point system and numerous rewards) created tangible evidence that youth could excel despite the overt and covert messages and practices associated with segregation.

Given the contemporary patterns of segregation and the predictable clustering of "at-risk" students, it is prudent to consider the value of creating counter-narratives today. Again community and full-service schools, when highly functioning, create environments where acknowledgements of student aspirations extend beyond lip service. Programming during and after school purposefully nurtures natural talents, absent a deficit-driven punitive approach.

The contemporary model of accountability in too many cases punishes the students most in need of additional resources. We do not have to look far for evidence of this claim (Orfield, 2009). Community schools by virtue of their mission direct more of their attention and resources (not always in the form of money) to services that might seem ancillary to the learning process, but consistently yield improved outcomes on standardized measures of academic progress. One of the reasons full-service schools are positioned to reverse tides of underperformance lies in operationalizing protective programming that results in resilience. By virtue of the model in its purest form, there is an acknowledgment that students do not leave the challenges they face outside the school. Second, there is recognition that unthwarted risk factors can directly and indirectly hamper optimal academic achievement and adult development. Third, schools are a rational site for the delivery of services because as opposed to the twentieth century, they are one of the last remaining dependable institutions in every community. Transportation issues that hinder access to a medical home for example are alleviated when health services can be offered in schools. More than words, our actions demonstrate to students what we believe about their potential. Investments of time, money, and other resources in a community school context hold the potential of creating counter-narratives that combat negative images and equip children with an internal confidence in their abilities, a protective buffer of sorts that speaks louder than any lecture.

Carter G. Woodson reminds us that the power of education lies in the power it tenders to its recipients because it increases social, economic, and cultural capital. He believed that power is dispensed and nurtured by educators when he said, "Real education means to inspire people to live more abundantly, to learn to begin with life as they find it and make it better" (Woodson, 1990/1933, p. 29). Evidence of the significant power welded by teachers can be found in historic glimpses of social change movements such as the Civil Rights struggle. African American teachers in particular were pivotal in dismantling educational and social inequities grounded in systematic governmental disenfranchisement of youth.

It unwise to talk about educational reform without considering the impact of teaching practices and curriculum, and it is additionally counterproductive to exclude from consideration the personal biases teachers bring with them into their classrooms and how they impact the educational prospects of students.

Teachers bring with them their own dispositions formulated by their social economic and ethnic classes as well as life experiences. These biases influence the delivery of educational services and a willingness (or lack thereof) to adopt practices designed to empower youth. Literature is replete with examples of how preconceived notions about children's behavior and potential often become self-fulfilling prophecies (America's Promise Alliance, 2008; Benard, 2004; Buttery & Anderson, 1999; Freire, 1998; Hilliard III, 2003; Jensen & Fraser, 2006; Reiss, Colbert, & Hebert, 2005; Richardson, 2008; Steele, 2003). To be fair, teachers are also subject to numerous forces that influence education delivery. Student achievement is also predicated upon inputs from parents, legislative bodies, education interest groups, education credentialing bodies, society's purpose of education, community organizations, and the business community. And we should not forget that educational attainment is also heavily influenced by children themselves who come to school with their own internal motivation. The tangled web of politics, teacher preparation and biases, and childhood challenges such as poverty are just a few strands of the web that predict the ability of educators to ignite the spark of learning.

I argue where the conversation about community schools needs to begin is with educators. Failing to oppose the standardized nature of schooling perpetuates an unfair and inefficient system for too many children. If we affirm that position, the next step is toward action. What interventions and pedagogical strategies are we willing to change to ensure an equitable educational experience for all of America's children? Teachers and principals in the African American tradition are used to the symbiotic relationship between professional responsibilities and advocacy opportunities and here we can take a lesson from JACS.

At its root, pressures relative to school accountability denote movement toward more top-down control of schools. It appears that public officials (either elected or appointed) believe promotion of student achievement can be legislated through sanctions, distribution of school data (e.g., test scores), student demographic information, and teacher credentials (Moe, 2003). It is a bit of a bullying approach to improving what is clearly ineffective. Twenty years of standards-based reform has done little to obliterate the achievement gaps experienced by some children, which hints at the inadequacy of the law alone to insure outcome parity (Gregorian, 2004). Exemplifying the ineffectiveness of our current approaches consider that unlike comparably developed nations, our high school dropout rates in the 50 largest cities in the United States average 58 percent (Swanson, 2008).

To create islands of opportunity for African American children in particular and other disenfranchised youth by inference requires an honoring and application of historically effective practices. In a post – Civil Rights era there are a number of potential JACS lessons. Public educational settings are not intentionally organized to nurture African American students as natural achievers. If students demonstrate prowess in subjects like mathematics, or the sciences for example, they are viewed as exceptional or "special" as opposed to beginning with the assumption that within all African American children (and all children for that matter) resides the potential to be world class scientists, economists, and mathematicians. Some of this failure to acknowledge the potential of youth is embedded in racist ideas of intellectual capacity. Failure to name racism as a real threat to children perpetuates its subversive power to undermine academic achievement (Hilliard III, 2003).

For individuals outside the experience of African Americans, it might be puzzling to propose that the subject of racism is still germane in a year when the nation elected its first African American president. While certainly evidence of the evolved thinking of many individuals, disenfranchising practices still exist in schools across the country whether purposeful or rooted in benign neglect. Bear in mind that for no other sect of the national population has there been more attention given to unabated ideology about intellectual competence (Perry, 2003).

It is past time for an alternative conceptualization of intellect. Consider intellect as expandable through effort and experiences. Yes, one is born with certain natural abilities, but giftedness in any discipline benefits from targeted support. Steele's (2003) extensive body of research revealed that just as stereotype threat (which would be fostered by either racism or lowered expectations fueled by other educators' ideologies) can perpetuate a manifestation of lowered performance, conversely, affirmations and high expectations prompt students to meet or exceed the norm-referenced performance.

To be fair, there are well-meaning professionals who feel overwhelmed by the prospect of remediating hindrances to learning. In this context lowering standards might appear to be a most humane and thoughtful approach. However, not only do lowered expectations send damaging messages to children that may be absorbed into young psyche but the resiliency literature and examples of highly effective schools in neighborhoods plagued with "risk factors" speak to the power of maintaining high expectations for academic achievement.

Exorbitant attention has been paid to risk, which is a decoy issue (Hilliard III, 2003). By that I mean as long as educators are focused on factors over which they either have no control or choose not to mitigate, attention is diverted from doing what is necessary to promote student success. Students thirst for venues to demonstrate their full potential. Programming can be those platforms of protective influence instead of the counterproductive decoys of "teaching to the test." We see in the accounts of JACS that commitment to fine arts, recreation, and vocational skills helped close the opportunity gap. Likewise we should not sacrifice well-rounded curriculum for test-prep because in- and after-school activities inform and reinforce one another.

For purposes of furthering the validity of a counter-narrative, consider the ways in which community school personnel can take on the persona of a resiliency partner with families. Harkening back to the JACS contributors to a successful community school, contemporary personnel and partners must figure a way to deliberately affirm and demonstratively honor the social, cultural, racial, and ethnic identities of pupils, all the while insisting upon rigorous curricula. Construction of programming to meet needs external to the school life and weaving academic skill building is an art whose merit is revealed in both historical and contemporary community schools.

It should be noted here that I do not ascribe to the notion that to teach students of color one must be a member of the aforementioned group. What I am suggesting is that the community and full-service school model when constructed according to its ideals honors the traditions, mores, and assets of respective neighborhoods, thereby enhancing the likelihood that the school will be a gap closer.

One unanticipated finding was verification all of the JACS teachers were reassigned to White schools post-*Brown*. Unlike Charles Johnson's (1954) prediction that integration would displace, demote, and decrease the numbers of viably employed Black teachers (which was the case in many if not most communities), JACS teachers were not among that number. There were most likely several reasons for this to happen. First, the relationships the Andersons had established over their long tenure; the highly visible nature of the school; and the academic prowess of students (and by inference the teachers') would make retaining them in the district an attractive option. Second, they were some of the best-prepared academics in the city. Third, the school system had only been partially segregated and teachers had attended district-wide staff meetings together. Most, it should be noted, were assigned to the elementary schools; with the exception of Mrs. Anderson who became the first African American teacher at Scott High School.

It would be wonderful to set forth a one-size-fits-all advocacy strategy for community schools that could be used by teachers and other school personnel. However, as has been noted several times, each school community is subject to its own policies, political actors, assets, and deficits. What is clear is that if we as educators do not attempt to address the shortcomings of traditional educational delivery we will perpetuate achievement gaps of growing numbers of students.

Thus we flash back to the theoretical framework that guided this work and extract useful lessons from JACS in the major categories of leadership, partnerships, organizational capacity, and benefits.

Applying the tenets of grounded theory methodology, the Highly Effective Community School (HECS) model (Figure 7.1) was created to both explain categorical contributions of JACS and advance a holistic conceptualization of community schools. Chapters 4, 5, and 6 deconstructed the

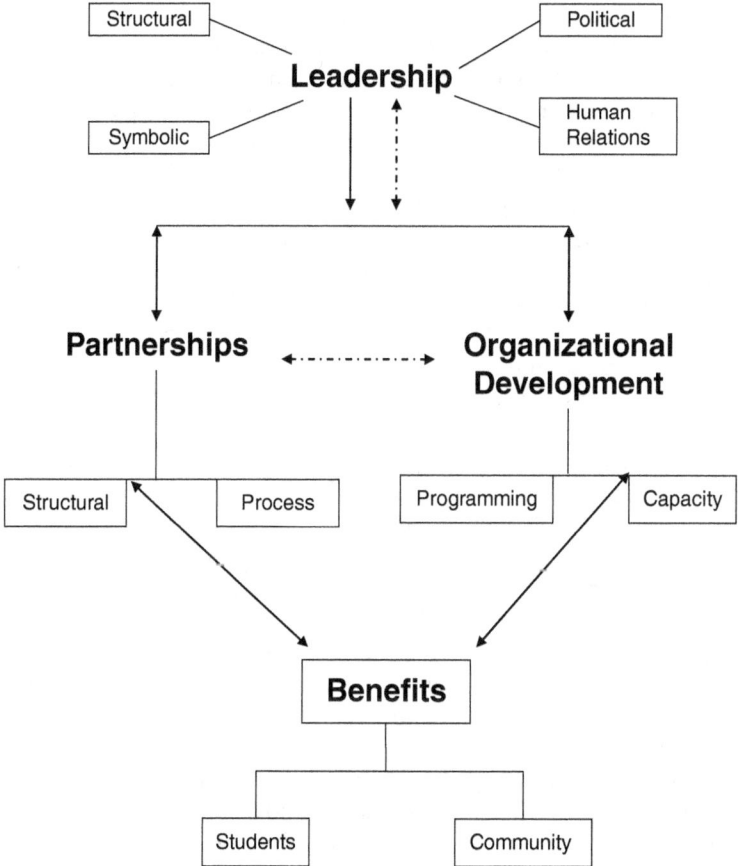

**Figure 7.1** Highly Effective Community School (HECS) Model Revisited.

HECS model associating documented evidence of practice. Now let us reconstruct it and extract the contemporary lessons to be learned and how dimensions of JACS might be replicated.

### Lessons in Leadership

Classifications comprising leadership in the HECS model are structural, human relations, political, and symbolic (Rice & Harris, 2003). In this school and others where leadership catapulted schools into highly effective community schools certain themes emerged. Principals operationalize the notion that their relationships with partners are paramount to the success of the work and those relationships are nurtured over time (human relations). Educators should conceive networks that support the advancement of learning and draw upon the skills, knowledge, and resources of the broadly defined community (structural and political frames). External support of the organization serves to mobilize internal support and vice versa. Community voice in setting agenda, programming, etc., is critical to ensure meaningful engagement of partners (human relations and political). Space in the curriculum and after-school programming for professional development opportunities should be immediately applied (human resource and structural frames). Power should be shared so that members of the neighborhood, parents, and teachers buy into and co-create the vision for the school (Rice & Harris, 2003). In the absence of sustained, systematic protection of creativity, leading by example, and openness to evolving priorities, an engaged environment cannon be sustained over time (symbolic).

Striking about this leadership dissection is that in essence JACS had the benefit of two innovative principals for the price of one principal and one teacher salary. Discrete references to either one were exceptionally rare in interviews. While this may not be possible to replicate in school settings today certain aspects of their united approach still apply. Principals often have assistant principals who if similarly committed and engaged can duplicate the multiplier effect T. J. and Anita created.

Structural contributions to leadership quality speak to the preparations of leaders, their expertise, and leadership styles. Effective leaders in traditional and full-service schools need to understand community challenges. There is no substitute for exemplary academic preparation, because it provides the basic skills needed to conceive a school curriculum inextricably bound to "real life." This alone is insufficient if not paired with a degree of cultural humility that is never patronizing of families, students, and partners. An ability to adjust leadership style to divergent situations is also vital. The Andersons possessed between them two bachelor's degrees, three master's degrees, and a law degree. In addition both had experience as teachers and as principals. This combined pedigree provided pedagogical and curricular expertise, an understanding of the politics of schooling, and the pressures associated with inequitable funding of segregated schools. Knowledge of the Coatesville community was garnered in several ways. First, the Andersons invited community members into the school to share their concerns. Second, civic engagement in their personal lives provided insight into the issues facing Blacks in Coatesville. Their malleable leadership style allowed for delegation of programs and curricula that endeared them to faculty and others. Teachers felt that Mr. Anderson buffered them from the politics of the city council and school board, allowing them more time to focus on innovative interventions for children. Furthermore, because decision making was not exclusively authoritarian (or top-down), collaborators were empowered to construct the blueprint of their engagement. A healthy full-service school organization cannot be created singularly by the leader/principal. By delegating power, portions of the implementation can be carried by others, increasing what can be accomplished.

Accountability standards might tempt contemporary leaders to adopt more of an authoritarian top-down leadership approach for understandable reasons. However, the community school data as well as the vivid examples provided from this examination of JACS indicate that if one is willing to take a risk (symbolic) and loosen the reigns of control a bit, positive outcomes can result. When one first arrives at a school, be mindful that within each school there are most likely multiple cultures present, each possessing time-honored traditions. It is in the assimilation of idiosyncratic community needs into the curricula that children are equipped to fight stereotype threats and acquire internal counter-narratives. Seek advice about community assets, needs, and challenges from diverse sources as a first step of building the blueprint for your community school.

Human relations or the building of sustainable relationships is another quality of an effective leader. As a carryover from seeking information about your service area, one can begin to identify the organizations and persons best positioned to aid in the creation of a community school. The adage of garnering the support from "low hanging fruit," or individuals and organizations most vested in school success, is a good strategy demonstrated frequently by the Andersons. However, it does not negate the need to simultaneously seek political, corporate, and community allies.

Anita and T. J. tapped their personal affiliations (clubs, classmates, and communities of faith) for school support. They also sought unlikely colleagues. For example, White business owners, congregations, and political actors proved to be very helpful in the sustainability of the community offerings at JACS. Strategic alliances with colleges accrued to students (Cheyney State Teachers College and Lincoln University) in terms of exposure to opportunities. Some partners were targeted because they had a vested interest in outcomes that extended beyond the school. For example, keeping children healthy had implications for working parents (they would not miss as many days of work) and reduced the likelihood diseases would spread through families and beyond. Physicians and dentists were clearly vested in ensuring the health and well-being of residents (such as Drs. Atkinson and Howard).

A great lesson as it pertains to partners was the penchant to identify the nexus between school need, partners' mutual benefit, and gratitude for aid of any kind. Investments signaled to the Black community one's willingness to support the education of their children. Collaborations were acknowledged with awards, testimonials, and other forms of gratification that reinforced appreciation for partner investment.

In a contemporary context, a community school principal should develop both a comfort level engaging people of different backgrounds and a keen eye for potential contributors. Remember that nonacademic affiliations can be great sources of support. Teachers and staff are your partners in this work because you surely cannot implement the community school approach without them. As you acknowledge the contributions of others, loyalty to your leadership and to the vision is amassed. From the Andersons we can also learn that valuable expertise is not always apparent in superficial assessments of staff, faculty, and community members. If you begin seeking community assets you may be surprised how many persons will be willing to support the school if they are asked. This of course requires educator to forgo their separateness from the communities in which they are embedded. Defying professional turf refers to open consideration of potential partners who in a conventional sense might be ignored. Skilled seamstresses or plumbers may not possess college degrees, but they have much to contribute to the life of a school. Professional snobbery had no place in the effective organization spearheaded by the Andersons.

There is a certain vulnerability leaders need to embrace in their dealings with staff and potential partners. Possibilities exist that pleas for assistance and input will be denied initially. However, the Andersons teach us that leading by example (doing as much as you ask others to do) creates an aura of openness and authenticity that over time can win over naysayers. Their examples also instruct us as to the magnetic effect of success. As students become more engaged and proficient, the spillover benefits of school affiliation become more apparent, thus attracting support from previously reluctant quarters.

It is not enough to evaluate community assets and schedule appointments. Principals or their designees need to be adept at communicating with different audiences. T. J. and Anita were great messengers of the JACS mission. Today, with diverse cultures and ethnicities represented in school bodies, careful consideration should be paid to who should be the metaphorical face of the school when soliciting aid. At times a parent, teacher, student, or community activist might be the best advocate for your cause because their relationship pathways are more reliable than your own.

Educators may loathe talking about the political nature of their positions; however, the fact remains that elected and appointed bodies determine the scope of building level authority, the distribution of public dollars, and in some cases even the curriculum taught in classrooms. Schools in addition to being subject to traditional political bodies are political themselves. Harkening back to our earlier definition of politics, decisions teachers and principals make every day determine which children will be educated in ways that nurture innate intellect and talent.

Taking this line of thought a step further, it is not enough to know who the power brokers are in a (political) community. Full-service school leaders need to convince power brokers and political actors of the value of sharing their expertise and clout to advance the mission of the school. Thus, political prowess in the context of leader behaviors refers to the ability of principals and their designees to not only identify strategic partners (human relations) but then convince them to advocate on behalf of the school. A politically astute leader shares information in ways that gives the operation of the school and its partners a level of transparency. Resources are distributed based upon the greatest needs as cooperatively defined by the school and community.

Sharing of information about JACS investment allocations took numerous forms. Sponsored programs regularly included a list of patrons. *The Adams Gazette* and *Adamite* posted school calendars and event recaps. In addition to school-based publications, the *Concern*, the Interracial Committee's publication, and *The Coatesville Record* were used to advertise activities of interest to the community at large. These publications also served another function; they kept JACS in the public eye as a vibrant community-centered entity, which added to Mr. Anderson's largesse when he approached the school board and Coatesville establishments for support.

The *Concern* and *The Adams Gazette* were also employed as a type of bully pulpit to highlight the value of the community school model and integration. An impressive balance was struck by the publication, which could have taken a militant bent. Instead, the goal of articles was primarily to highlight examples of progress and provide tools, vocabulary, and advice to White individuals interested in more inclusive practices. Featuring inclusive practices in the absence of blame set the tone for a diverse constellation of parties open to support the mission of the school.

The Andersons were constrained by *de jure* segregation practices and contemporary schools are impacted by *de facto* segregation, yet both function with similar political limitations. First, the schools these leaders serve tend to be under-resourced relative to the needs of their students. Second, while they can lobby for additional public dollars this is a fight they are apt to lose. Leaders in this situation do well to take stock of community assets to identify the community-based power brokers (internal and external to the school) who might have a vested interest in supporting the community school in some way.

Another lesson we can extract from the Andersons is the value of powerful allies who based on their position can advocate on the schools' behalf. Two of T. J.'s champions were superintendents who openly opposed segregation and were committed to ensuring the Black children of Coatesville were well equipped to enter the integrated high school. Other allies include power brokers invested by virtue of their presence at school functions and being patrons of events (see Charles Huston, Jr., President of Lukens Steel Company). Persons of this ilk also include clergy (such as Rev. James Brown), who became catalysts for broadening White partnerships within JACS. Defying professional turf refers to open consideration of potential partners who in a conventional sense might be ignored. Skilled seamstresses or plumbers may not possess college degrees, but they have much to contribute to the life of a school. Professional snobbery had no place in the effective organization spearheaded by the Andersons.

Symbolic attributes of effective leaders are in part innate and in other respects behaviors and strategies to be learned. Crafting a guiding vision that is broad enough to be resonant with diverse constituency groups yet specific enough to be attainable is a difficult exercise. First birthed in the mind of the leader, s/he then needs to garner support in whatever way makes the most sense for the neighborhood served. The charisma to engage persons to embark on the collaborative journey of creating and sustaining a community school cannot be taught. However, the one primary reason the Andersons were so successful is articulated repeatedly by students and faculty and is evident in the many archival documents. They were personally and demonstratively committed to the success of the school and did not ask of anyone what they were not willing to do themselves. Authenticity cannot be faked. Children, their parents, and members of the community intuitively know if you are serious about inviting their engagement or not. They know if you sincerely believe their children have untapped potential, and are willing to work side by side to create a full-service school.

What I am proposing is labor intensive; however, leaders should not be intimidated by the enormous task of creating a full-service school. Identify messengers and allies early and some of the planning and implementation can be delegated. Another lesson from the Andersons was their incremental progress toward the full complement of services and activities. Numbers of classes increased over time and simultaneously they were willing to drop some offerings if attendance was lackluster. As a result, replication of the many after-school activities offered at JACS is not the goal. Instead, innovations need to be grounded in idiosyncratic school district priorities, assets, and mores.

Pictorially, the HECS model reflects a linear relationship between leadership, partnership, organizational development, and benefits. However, first prompted by the leadership literature and then findings of this study, it becomes overwhelmingly apparent that in the absence of strong leadership a sustainable highly effective community school is not possible. It is the principal who provides and protects the creative pedagogical space of teachers from outside sources. It is the leader who sets the tone for an environment supportive of high academic standards. The principal doesn't do the work of teaching and learning, but provides the canvas that teachers, students, parents, and the community collaboratively design. Sometimes this requires authoritative decisions and other times the democratic process works best. Security in one's strengths and a willingness to seek assistance seem to be part of the formula for success. Second to leadership are strategic partnerships that enable the work of community schools to be accomplished.

### Lessons about Partnerships

It is not reasonable to assume that school faculty and staff alone can create the complex web of services needed to convert a traditional school into a full-service community school. Comprehensive services and activities need to be school linked or in the schools, and undergirded by their own funding. Chapter 1 articulated the likelihood that the children who would most benefit most from this model of schooling are most apt to attend under-resourced schools, that is, majority-minority school populations segregated in urban enclaves. As contemporary budgets are constrained at least in part by costs associated with standardized testing, T. J. and Anita Anderson's resourcefulness provides instruction on how to encourage community coalescence around the full-service concept with minimal resources. Thus, given the strain on school budgets, lessons from a similarly situated community school provide helpful ideas.

Strategic partnerships are one way to bridge the gap between needed services and constrained budgets. Sources range from community-based organizations, clubs, and businesses, to parents, teachers, and staff of the school. Anyone willing to add to the life of JACS was welcomed and a place made for their unique contributions. As noted in Chapter 5, associates might have contributed a one-time gift or extended commitments of time and resources over multiple years. Consider the many nationally renowned visitors to the school. Their contribution might have been one episode, but their influence lasted much longer than their visits. Students saw what might be possible for themselves in the operatic voices of Leontyne Price and Roland Hayes. They too could excel in sports that had historically excluded them, like Jackie Robinson and Walter Triplett. College presidencies were within their grasp after meeting Dr. Leslie Pinckney-Hill. The point is contributions were welcomed as long as they supported the mission of the school.

Partnerships as per the model can be divided into structural (who the partners are) and process (what the partners do and how they function with relation to the school). Though pre-dating the cited literature, JACS partners included the prescribed gamut of players. Teachers, parents local businesses, the faith and medical communities, civic clubs, and advocacy organizations representative of White and Black organizations provided varying levels of support to the school. The lesson to be learned here is to cast your net broadly and allow some flexibility in the participation level of partners.

Participation levels of all types and sorts (monetary, time, expertise, or public relations) are valuable in establishing a community school. Reflecting back on JACS, partners embracing the school and social ethos of CIHU remained connected the longest. It is worth pausing here for a moment to reflect on why that might have been the case in an effort to replicate this phenomenon. There seem to be two major reasons collaboratives lasted at JACS. First, engagement was mutually beneficial. For example, advertising in the school newspaper might increase store revenue or there might have been a common goal of racial equality as was the case with the YMCA and JACS. Second, once the level of partnership was established, there were overt and subtle appreciations rendered, demonstrating not only gratitude but the quantifiable value contributions of any kind added to the JACS community.

From the cross section of JACS partners we are instructed in the ways the Andersons began the process of creating a community school. They began in the building. It will be difficult to garner support for your programs if the building-level personnel are not clearly vested. They are vital partners and because of their intensive interaction with students will be able to provide insight into the community assets and power brokers. When seeking partners do your homework to determine what student or community benefit might best align with prospective collaborators' mission, goals, or objectives. Next, allow partners to dictate the level of their engagement and remain somewhat flexible. It is better to start with small commitments of time and money, allowing time to assess the return on investment. Then informed decisions can be made to become a long-term partner. That said, one-time donations can be valuable but do not add to the long-term sustainability of the community school.

Faculty were invaluable in the mid-1940s, because support for after-school initiatives could not be staffed by community volunteers and three paid staff alone. The teachers were significantly engaged in life after school without additional compensation. As the notoriety of the JACS outcomes became known, more resources supported additional paid after-school staff. However, even after they could presumably have left the oversight to others, teacher remained substantively engaged in after-school programming as volunteers.

It is not as likely that the time commitment of teachers in particular could be duplicated today for a number of reasons. As a pragmatic matter, teachers rarely live in the same communities in which they teach and as a result have to travel to care for their own families. That said it will take more volunteers from the community and other sources to support after-school programming. In some school districts there are funds available to support after-school academic remediation that could subsidize teachers remaining after traditional school hours. Bear in mind that after-school remedial classes alone do not constitute a community school.

Partnership processes speak to what is actually done and how interactions are woven into the fabric of school life. Opportunities for contributions abound in community schools, for example, students in need of mentors or tutoring. Businesses can allow students to shadow professionals and provide financial support. An important element in the partnership process is the latitude to make suggestions relative to the programmatic offerings and how suggestions will be implemented.

## Lessons in Organizational Development

Community schools are organizations, and organizations are administrative or functioning structures or social units of people systematically arranged and managed to meet a need or to pursue collective goals on a continuing basis. Given this definition as a starting point, all schools exist to promote the learning of youngsters and prepare them to become contributing members of society. Their governance structures and funding formulas vary from state to state and city to city, but their rhetorical missions remain the same whether the school functions traditionally or as a community school.

Full-service schools differ from their traditional counterparts in one significant way. Children are expected to adjust to the rules and regulations of traditional schools that function with a goal of standardization. Conversely, community schools adjust to meet the needs of the child. Ultimately both institutions seek a similar final end product (high academic achievement) but go about attaining this goal in very different ways. Since a community school's approach to learning is more holistic, assessments of community assets and needs are the first step in identifying potential programming and the type of organizational characteristics that will be needed for implementation. Capacity then comes into play as decision makers determine the organization's ability to commandeer the resources to sustain programming.

As has been noted throughout this book, adherence to the truest form of the community school model would ensure the community informs and supports the after-school programming. Meaningful engagement of communities while laying the foundation for a community school requires a scaffolding approach to organizational development. Building personnel need to be well qualified to advance subject matter competencies and clearly articulate performance expectations. High expectations, demonstrating a belief in students' ability to meet performance expectations, are also important. Professional preparations aide in the construction or adaptation of curriculum that is relevant to students. As John Dewey (2001/1915) would attest, the most successful community schools are those where there is no clear demarcation between the school day and the material covered and after-school activities.

Opportunities to practice and apply core subject knowledge at JACS were purposeful and did not ignore the importance of incorporating a cadre of recreational activities. What a glance at the list does not tell us are the many ways content competencies were woven in ways the students barely noticed. For example, selections tied to the Evelyn Lawrence Oratorical Contest drew from history, English, and social studies curricula. The school bank and woodshop classes required participants to apply mathematical skills. Spring musicals and operettas brought ancient texts to life and band members playing for musicals had incentive to practice. Another example was the weaving of healthy lifestyle choices and the need for exercise throughout the physical education and home economics curricula and in May Day exercises.

Evening offerings also stressed the building of character and connection to the community under the umbrella of CIHU. Boy and Girl Scout troupes and the ways of earning the badges aligned with CIHU activities such as the clothes closet and food pantry, as did the donation of arts and crafts work to those in need. Interspersed in the already busy JACS evening program were numerous vehicles where students, parents, and community members were thanked for their contributions. For students, Move-Up Day, graduation ceremonies, and achievement banquets encouraged them to strive for excellence. These forums, as well as others, were also the time that community members were honored for outstanding achievements or thanked for their participation with the JACS community.

We learn from this litany of activities that as an outgrowth of community input, afternoon and evening activities can excite adults and children alike and enhance the sense that they are vested in the school. Enthusiasm is contagious and will support your efforts as you seek to expand program offerings. Think carefully about which activities to continue and which to discontinue. If attendance is low and priorities shift be prepared to make adjustments. Community needs are fluid and programming should be responsive to shifting priorities.

Some things that did not appear to be issues with JACS can present significant hindrances to after-school programming now. Safe and affordable transportation to and from the school is easier to manage during traditional school hours. Risks to children traveling alone have increased exponentially. This is not an impossible challenge to overcome, if one adapts the "school as community hub for problem solving" mentality. Seek advice from your community partners because they may have creative solutions, such as walking/traveling groups. Issues of insurance coverage during off-peak hours may also be an issue in the creation of a community school. Your district will have individuals with the expertise to advise you on your limitations.

As was noted earlier, organizational capacity refers to the ability to sustain the programming envisioned by the school – community alliance. Critical to capacity is an assessment of the physical plant. The school building attributes determine not only what programming can be offered, but also how many different activities might be supported simultaneously. Next one should consider what contribution partners can make within the confines of your space. For example, can you accommodate a school-based health clinic in the building that would offer more comprehensive care than the district allows school nurses to provide or is the community health clinic close enough to be considered school linked? How much money is needed to cover not only the cost of each activity, but also to include support staff, increased insurance, utilities, and so on? Once you have a sense of the costs, you can go about seeking sources for those funds. Given contemporary budget constraints, school boards may not be in a position to allocate or unwilling to divert resources to support your cause.

Here is where the tenacity of the JACS administration and staff is most informative. It appears they left no potential donor untapped. Though classes and activities were offered free of charge, donations were sought regularly. The annual winter carnival was the major fund-raiser for the evening program. The YMCA partnered with JACS both in terms of financial resources, and volunteers. The PTA fund-raisers helped with not only traditional school needs, but also after-school expenses. Programs such as the operettas, plays, and other concerts were at least partially underwritten by individual and corporate patrons. As impressive as fund-raising prowess was, it would still not have been sufficient to keep the evening classes going without the investments of volunteers and in-kind donations that came from students, faculty, staff, parents, and community members, some of whom did not even have children in the school.

Perpetual fund-raising might seem an arduous task; however we can learn some lessons from JACS in this regard as well. First, it appears that because contributing to JACS was deemed a wise investment, it did not require what has been called a "hard sell." Second, the commitment of the school to be co-created with the community meant that in times when JACS needed more the community sacrificed to provide and when community members were in need, the school sacrificed to fill the gap. Reciprocity added to the cache of the organization. Ultimately, for a community school to be sustainable, tangible and intangible benefits must accrue to students and the community (to include partners).

## Lessons of Student and Community Benefits

One of the most poignant lessons we discover from JACS is that while the academic curriculum possessed sufficient rigor just raising standards was not enough. Children were provided a safe environment where they could seek assistance and explore interests. Civic engagement in the form of CIHU created added motivation to do well in school beyond the next test or marking period. It was tethered to honoring those who had invested in you. Students were connected to caring adults beyond their families who had demonstrated caring not with words or lectures alone, but by virtue of spending time and resources. Deconstructing "decoy" walls between schools and communities provided opportunities to create counter-narratives opposing negative images young people saw of themselves in media and by virtue of day-to-day interactions.

For the neediest of students (neediest here refers to youth exposed to multiple risk factors) school is the one reliable place they will convene, providing us opportunities to plant seeds of hope. This is a bit of an oversimplification of this comprehensive evaluation, but some features of school boards, students, parents, teachers, and communities have not changed in 60 years. School board members care about children but they also tend to be elected officials who are judged by school test scores, graduation rates, public image, and fiscal responsibility. Parents want their children to be safe en route and in school. They want youngsters to be engaged in activities that will identify their gifts and areas in need of support. In short, parents want their children's potential nurtured. Teachers want to work in safety with space to be creative. They want support from their administrators that ensure classrooms are places to learn and not holding facilities. They want parents to reinforce the value of academic achievement and children to come to school ready to learn. Students need and crave caring adults in their lives that ensure their safety and nurture their resilience. They want to be challenged to work hard so that accomplishments earned are valued and linked to a work ethic that will serve them well in life. Members of the community want schools to produce citizens who will contribute positively to society. In the meantime, they want to see that tax dollars are being well spent by virtue of valued outcomes and would welcome outreaches that were grounded in genuine concern for community well-being.

Outcomes letting us know our community schools are effective can be divided into traditional (or tangible) and nontraditional (or intangible) categories. Examples of traditional outcomes are graduation, attendance rates, and test scores. More difficult to quantify are nontraditional or intangible benefits of being part of a full-service school that cascade out into communities yielding advantages. Students profit because a safety net (supportive ecologies) has been created for them and as a result they are more apt to realize their potential, which includes, but is not limited to, academic achievement. Parents benefit because the burden of worrying about whether certain needs are met has ameliorated and because of the welcoming environment at the school they feel part of that community. As their schedules permit they are then more likely to contribute their time and at the very least partner with teachers to ensure consistent reinforcement to students. As was the case with JACS, parents have their own sets of needs which if unmet trickle into the lives of their children. As a result parents and other adults can benefit from after-school programming.

Faculty are charged with many responsibilities that extend beyond teaching. A major benefit of a community school is that teachers lend aid and receive it as well. Extra hands, eyes, and resources free them to concentrate on innovative pedagogical practices and rigorous curriculum.

A word here about the value of academic rigor. JACS's curricula were actually more comprehensive than those of its White counterparts, which equated to students being equipped to function in the integrated high school. The many opportunities they had to practice their skills built a confidence in their ability to perform anywhere. They were secure in their intellect because of a curriculum that wove the heroes, scholars, poets, and artists of the African American tradition seamlessly with the likes of Thoreau, Longfellow, and Dickens. As was noted earlier the teachers were among the most educated in the district adding the value of their preparation to the curriculum. Though Mr. Anderson was first recruited to Coatesville because of the abysmal state of Black education in Coatesville, by the time he retired, JACS was responsible for marked improvement in attendance, graduation rates, and college matriculation in the Black community.

Regularly honoring community members and faculty for their hard work seems to be one reason people stayed connected to the school (both a tangible and intangible benefit of participations with JACS). Hard work was acknowledged publicly and valued. However the greatest nontraditional benefit of attending JACS was the infusion of the CIHU ethos into the lifelong practices of students. In interview after interview and survey after survey, the commitment to civic engagement and helping others was articulated as a lifelong practice as demonstrated in Chapter 6 vignettes.

### Children Are Not Widgets: Looking Backward to Plan Ahead

A widget is a term most often used in economics literature to refer to a hypothetical product of a manufacturing. The argument is that the manufacturing process can be explained and evaluated in the various segments of production, for example, the purchasing of materials; the making of the widget; the marketing of the widget; and, finally, the profit or loss associated with the sale of the widget. I bring up this apparent departure from our educational focus because the language of business has increasingly trickled into educational policy and practice. One could argue that the accountability standards and punitive consequences associated with not attaining benchmarks for "success" has a bit of a corporate ring. However, children are not widgets. Educators do not control all contributions (or deterrents) to learning as a factory owner controls an assembly line. Try as we might to isolate the learning and achievement process to the exchange between teacher and student during the school day, volumes of literature explain not only that this focus has not worked, but worse, it has successfully perpetuated opportunity and achievement gaps.

Full-service community schools at their best create a "world" safe from negative predictions where students are affirmed and nurtured. JACS was that haven of hope for youth who without it would have traversed their educational pathways with great difficulty, if at all. Extracting lessons espoused by the composite leadership profile, partnership, and organizational development strategies of Adams, we see that once success takes root, children are better equipped to go out into a world that ascribes their risk as irrefutable and by tapping into their counter-narrative defy predictions tied to their race, class, or ethnicity.

Just as children are not widgets, community schools are not a panacea, nor are they identical. Each one offers a different scope of services and programs because of the community-specific resources and limitations of each school. While, I believe, it is possible for all children to benefit from attending a community school, some children would benefit more than others. Deciding to pursue the creation of a full-service school will require political skill. There are at least two ways to fight the status quo that are not mutually exclusive. One is to co-create with communities learning environments that cater to children and ask as permission to do so later. Another is to commit oneself to advocating for this approach by reminding political actors of the high price of benign neglect. The teachers at JACS and the Andersons did both and in the wake of "No Child Left Behind" and the impending reauthorization of the Elementary and Secondary Education Act it is time for educators today to do the same.

Educational apartheid may sound harsh as a label for the stubborn educational inequities that persist when a child's race and income converge. However, I am not the first nor, I suspect, the last to draw the inference. Educational attainment is still the greatest predictor of upward social and economic mobility. It has been and continues to be the gatekeeper to adult opportunity and without it citizens are doomed to piecing together resources for survival. Pressures to prepare students to compete in a global economy notwithstanding, schools increasingly find themselves sandwiched between the multiplied disadvantage experienced by disproportionate numbers of minority children who are also poor and calls for high academic outcomes measurable almost exclusively by standardized tests. What these trends portend is a growing number of students who will not have access to educational environments tailored to ensure success.

If educators, politicians, and the general public resign themselves to the inevitability of tiered access to quality educational experiences, we are in fact complicit in perpetuating educational apartheid for certain subgroups of the nation's children (Cool, 2005). If, however, we loosen our grip on uniform school practices and consider alternative approaches such as those advanced by the JACS, we at the very least have abandoned what we know is not working. Imagine, "A community school . . . established to perform [the] job of total education. . . . An integrated day and night program [that] caters to all elements of the population and seeks to bridge cultural gaps which are largely the result of unfulfilled group needs. The ultimate objective of the programming is to develop good American citizens who will make positive contributions to society" (Anderson, November, 1953). If it was possible in 1953 when T. J. Anderson penned these words, I am convinced that it is possible to create responsive community schools in the twenty-first century.

## References

America's Promise Alliance. (2008). *Every child every promise: Turning failure into action*. Retrieved February 2, 2009, from http://www.americaspromise.org/uploadedFiles/AmericasPromiseAlliance/Every_Child_Every_Promise/ECEP_Reports_-_JPEG/ECEP%20-%20Full%20Report.pdf.

Anderson, T. J. (November, 1953). *Coatesville City School District minutes*. Coatesville: Chester County Historical Society Archives, Coatesville City School District.

Benard, B. (2004). *Resiliency: What we have learned*. San Francisco: WestEd.

Buttery, T. J., & Anderson, P. J. (1999). Community, school, and parent dynamics: A synthesis of literature and activities. *Teacher Education Quarterly, 26*(4), 111.

Freire, P. (1998). *Teachers as cultural workers: Letters to those who dare teach*. Boulder, CO: Westview Press.

Gregorian, V. (2004). No more silver bullets: Let's fix teacher education. In J. Matthews (Ed.), *The last word* (pp. 10–16). San Francisco: Jossey-Bass.

Hilliard, A., III, A. (2003). No mystery: Closing the achievement gap between Africans and excellence. In T. Perry, C. Steele, & A. G. Hilliard III (Eds.), *Young gifted and Black: Promoting high achievement among African-American students* (pp. 131–166). Boston: Beacon Press.

Jensen, J. M., & Fraser, M. W. (2006). A risk and resilience framework for child, youth, and family policy. In J. M. Jensen & M. W. Fraser (Eds.), *Social policy for children & families: A risk and resilience perspective* (pp. 1–18). Thousand Oaks, CA: Sage Publications.

Kozol, J. (2005). *The shame of the nation: The restoration of apartheid schooling in America*. New York: Crown.

Moe, T. M. (2003). Politics, control, and the future of school accountability. In P. E. Peterson & M. R. West (Eds.), *No Child Left Behind? The politics and practice of school accountability* (pp. 80–106). Washington, DC: The Brookings Institution.

Morris, J. E. (2004). Can anything good come from Nazareth? Race, class, and African American schooling and community in the urban South and Midwest. *American Educational Research Journal, 41*(1), 69–112.

Orfield, G. (2009). *Reviving the goal of an integrated society: A 21st century challenge.* Los Angeles: The Civil Rights Project/Proyecto Derechos Civiles at UCLA.

Perry, T. (2003). Up from the parched earth: Toward a theory of African-American achievement. In T. Perry, C. Steele, & A. G. Hilliard III (Eds.), *Young gifted and Black: Promoting high achievement among African-American students* (pp. 1–108). Boston: Beacon Press.

Reiss, S. M., Colbert, R. D., & Hebert, T. P. (2005). Understanding resilience in diverse, talented students in an urban high school. *Roeper Review, 27*(2), 110–120.

Rice, D., & Harris, M. M. (2003). Leadership in community schools: A frame analysis. *Reclaiming Children and Youth, 11*(4), 216–211.

Richardson, J. W. (2008). From risk to resilience: Promoting school – health partnerships for children. *The International Journal of Educational Reform, 17*(1), 19–36.

Steele, C. (2003). Stereotype threat and African-American student achievement. In T. Perry, C. Steele, & A. G. Hilliard III (Eds.), *Young gifted and Black: Promoting high achievement among African-American students* (pp. 109–130). Boston: Beacon Press.

Swanson, C. B. (2008). *Cities in crisis.* Bethesda, MD: Editorial Projects in Education Research Center, supported by America's Promise Alliance and the Bill & Melinda Gates Foundation.

Woodson, C. G. (1990–33). *The mis-education of the Negro.* Trenton: Africa World Press.

# Index

Academic Outcomes, 1, 11–13, 16, 17, 19, 20, 27, 38–42, 47, 48, 52, 55, 61, 69, 117, 125, 127, 128, 131, 133, 148, 157, 158, 162–70, 175, 177, 178–83
African American(s)
Black(s), 2–6, 11–16, 24, 26–35, 39–41, 47, 48, 55, 60–3, 66–72, 74–9, 81, 82, 84, 89–94, 96–9, 101, 103, 104, 113, 115, 117, 119, 120, 126, 127, 129, 131, 143, 145, 146, 150–2, 155, 162, 163, 165–8, 171, 172, 174, 176, 182, 184
American Indians, 4, 11–13, 15, 29
Americanism Cup, 100, 122, 127
Anita Anderson, 2, 3, 7, 49, 69, 76, 87–99, 101–3, 106, 108, 111, 113, 117, 119, 120, 125, 128, 129, 133, 142–6, 148, 150–3, 155, 156, 158, 163, 168, 170–6, 183
Arne Duncan, 6

*Brown v. Board of Education*, 24, 27–29, 31–35, 41, 48, 59, 66, 69, 80, 117, 144, 156, 163, 168

Can I Help You (CIHU), 7, 80, 82, 83, 95, 96, 102, 103, 107, 121, 123–5, 127–9, 132–4, 140, 145, 147, 149–53, 156
Cecil Rice, 100, 120, 130
Characteristics of a good community school, 24, 25, 168

Charismatic Leader, 26, 34, 51, 101, 107, 108, 113
The Children's Aid Society, 2, 55, 56, 58
Coalition for Community Schools, 22–5, 32
Coatesville
Bridgetown, 2, 7, 32, 65, 69–77, 79–84, 89, 90, 92–5, 97–101, 103–6, 113, 115–17, 119, 120, 122, 125, 127–31, 140, 143, 144, 146–8, 150, 152, 154–7, 163, 171, 173, 174, 176, 179, 180, 182
Communitarian paradigm, 20
Community engagement
Community partnership(s), 17, 20, 22, 23, 25, 30, 38, 44, 54, 55, 68, 92, 94–6, 107, 108, 112–14, 119, 122, 131, 154, 157, 158, 170, 171, 174, 175–80, 182
Concern, 75–7, 82, 93, 142, 144, 173
Counter-narrative, 30, 33, 41, 48, 58, 82, 158, 162, 164, 165, 168, 171, 180, 183
Cultural competence, 52, 54, 163
Cultural humility, 55, 57, 170

David Turpeau
Ila Marshall, 90, 119, 129
Definitions of community and full-service schools, 18, 48, 53–5, 83, 177
Desegregation, 2, 27, 29, 79, 100, 117, 155

Dr. Whittier Clement Atkinson, 75, 96, 120, 121, 133, 155, 172

Ecology of Human Development
  EHD, microsystem, microsystems, mexosystem, mesosystems, exosystem, exosystems, macrosystem, macrosystems, 44–7, 49, 164
Eleanor Roosevelt, 4, 21, 74, 98
Elsie Clapp, 4, 6, 18, 20–2, 25, 26, 57, 66, 112

Faith communities
  Quaker, Presbyterian, Methodist, Baptist, Seventh Day Adventist, Catholic, Jewish, 54, 56, 66, 70–3, 75, 77, 79, 89–91, 96, 101, 102, 107, 113, 114, 118–21, 150, 151, 171, 176
Full-Service Community Schools, 2, 3, 6–8, 11, 16–19, 22, 24, 26, 29, 32, 38, 42–4, 47–51, 53–8, 66, 68, 69, 73, 78–84, 88, 89, 92–8, 104–8, 111–13, 115, 117–25, 131, 132, 134, 154, 156–8, 161, 162, 164–8

Gordon Junior High, 81–3, 100, 116
Guiding principles of JACS, 104

Highly Effective Community School Model
  HECS, 7, 49, 50, 56, 59, 84, 88, 94, 112, 121–2, 156–8, 169, 170, 175
High-poverty schools, 12–15, 17, 31, 39
Hispanic American
  Latin American, Latino, Latinos, Hispanics, 6, 12–16, 31
Howard University, 79, 80, 89–92, 96, 116, 150, 153
H. R. Vanderslice, 1, 79, 99, 120
Human Relations Leadership, 7, 52, 53, 58, 87, 88, 94–7, 107, 108, 169–71, 173

Hutchinson Memorial United African Methodist Episcopal Church, 73, 79

Individualistic Paradigm, 20
Integration, 20, 27–9, 76, 100, 104, 117, 120, 127, 144, 146, 155, 158, 168, 174, 182
Interracial Committee, 75–7, 82, 93, 98, 143, 145, 173

James Adams Community School
  JACS, James Adams, Merchant School, 2–8, 27, 29–32, 41, 42, 49, 52, 59, 66, 67, 69, 71, 76–84, 88–100, 102–6, 108, 111–34, 140, 143, 146–8, 150–8, 162–4, 166–84
James Morgan Adams, 79
Jim Crow laws, 29, 30, 66, 72
John Dewey, 3, 4, 6, 17–21, 44, 57, 91, 112, 157–9, 178
John Hope Franklin, 11
Joy G. Dryfoos, 1, 2, 17, 18, 20, 24, 25, 44, 53–5, 57, 58, 96, 107, 112, 114, 122, 124

Ku Klux Klan, 73

Leader behaviors, 25, 26, 50–4, 57, 58, 88, 107, 121–2, 163
Leadership, 4, 7, 8, 21, 25, 26, 49–54, 56–9, 84, 87–108, 112, 114–16, 121–2, 133, 134, 154, 156, 169–75, 183
Leontyne Price, 93, 99, 119, 176
Leslie Pinckney Hill, 99, 119, 129, 176
Lukens Steel Mill, 70, 71, 73, 74, 119, 129, 145, 152, 153, 155, 174

Majority-Minority Schools, 4, 13, 15, 17, 31, 175
Mary McLeod Bethune, 1, 98
Mason – Dixon Line, 66, 70–2, 78

No Child Left Behind
  NCLB, Elementary and Secondary
    Education Act, ESEA, 16, 24,
    30, 183

Oral Historiography, 3–5
Organizational capacity, 7, 25, 49, 50,
  56–9, 108, 111–12, 121–4, 156,
  169, 177–80

Parallelism, 78, 82, 106, 141, 145
Partnership
  partners, 1, 7, 8, 14, 17–19, 22, 23,
    25, 26, 42, 43, 47, 49–59, 80, 92,
    94–6, 100, 104, 107, 108,
    112–15, 117, 119–22, 132, 134,
    142, 156, 163, 168–70, 172–7,
    179–81, 183
Paulo Freire, 37
Political Leadership, 7, 46, 47, 50–4,
  58, 82, 87–90, 93, 97–100, 107–8,
  114, 115, 120–1, 169–71,
  173, 183
Poverty stressors
  poor health, poor housing,
    low-income, food insecurity, 13,
    14, 24, 27, 31, 39, 43, 69, 162,
    166, 183
Protective factors, 30, 41–4, 46–9, 55,
  157, 158, 164, 165, 168

Qualitative Research, 4, 5

Resilience, 2, 23, 39–44, 46–9, 55, 113,
  157, 162, 165, 167, 168, 181
Rev. James Brown, 71, 75, 96, 120, 121,
  130, 174
Risk and at-risk, 1, 13, 14, 17, 20, 23,
  24, 30, 31, 37–44, 46–8, 55, 140,
  157, 158, 162, 164, 165, 167,
  181, 183
Roland Hayes, 77, 99, 119, 176

Segregation
  Jim Crow, de facto, de jure,
    resegregation, 3, 14, 26–31, 41,
    47, 48, 50, 66, 68, 69, 72, 74,
    76–9, 81–3, 96, 97, 99, 100, 108,
    113, 120, 125, 144, 145, 154, 155,
    158, 162, 164, 168, 174, 175
Stereotype threat, 29, 39–41, 58, 68,
  157, 163–4, 167, 171
Structural Leadership, 7, 50, 52, 53, 55,
  58, 87–9, 97, 106–8, 111–14, 121,
  122, 169–70
Symbolic Leadership, 7, 50, 52, 53, 58,
  87, 88, 101–8, 133, 169–71

Thomas Jefferson Anderson
  Andy, 7, 8, 18, 29, 30, 49, 52, 68, 70,
    71, 75, 78–84, 87–108, 112, 113,
    115–22, 125, 128, 129, 132, 133,
    140–6, 148–53, 155, 156, 158,
    161, 163, 168, 170–6, 182–4
Transactional leaders, 26
Transformational leaders, 26

Underground Railroad, 70
University of Nottingham, 92, 112
Upper Octorara Presbyterian Church,
  71, 75, 120
Urie Bronfenbrenner, 44–6, 53, 68, 164

Wally Triplett, 99, 119
White American
  White, Whites, European American,
    Eurocentric, 12, 14 16, 27, 28,
    30, 40, 41, 66–8, 72–9, 81–2, 89,
    92–4, 96–101, 104, 107, 113,
    117, 119, 125, 126, 129, 131, 141,
    145–8, 150–3, 155, 162, 163,
    168, 171, 174, 176, 182

Young Men's Christian Association
  YMCA, 80, 95, 118, 124, 176, 180

Zack Walker, 72–3

GPSR Compliance
The European Union's (EU) General Product Safety Regulation (GPSR) is a set of rules that requires consumer products to be safe and our obligations to ensure this.

If you have any concerns about our products, you can contact us on

ProductSafety@springernature.com

In case Publisher is established outside the EU, the EU authorized representative is:

Springer Nature Customer Service Center GmbH
Europaplatz 3
69115 Heidelberg, Germany

www.ingramcontent.com/pod-product-compliance
Lightning Source LLC
LaVergne TN
LVHW011826060526
838200LV00053B/3917